SUSTAINABLE MASONRY CONSTRUCTION

Mark Key

 bre press

Fan-vaulted ceiling, King's College Chapel, University of Cambridge
Often referred to as 'the noblest ceiling in existence', it was built in just three years between 1512 and 1515 by master mason John Wastell

SUSTAINABLE MASONRY CONSTRUCTION

Mark Key

 bre press

Published by IHS BRE Press

IHS BRE Press publications are available from
www.brebookshop.com
or
IHS BRE Press
Willoughby Road
Bracknell RG12 8FB
Tel: 01344 328038
Fax: 01344 328005
Email: brepress@ihs.com

Printed on paper sourced from responsibly managed forests.

Cover images:
Main: Bath House, Barking Central project (© AHMM Digipic archive)
Top right: St Pancras Station, London (photo: Irvine-Whitlock)
Middle right: Hanson EcoHouse, BRE Innovation Park (photo: Nick Clarke)
Bottom right: Mariners Way, Whitehaven (photo: Ron Reed, High Grange Developments Ltd)
Back cover: Jerwood Centre, Grasmere, Cumbria (author's photo)

Prelim images:
Page ii: Fan-vaulted ceiling, King's College Chapel, University of Cambridge (author's photo)
Page x: Stanbrook Abbey, Yorkshire (photo: Peter Cook)
Page xvi: The Hothouse, Hackney (photo: Nick Guttridge; architect: Ash Sakula Architects)

Index compiled by Margaret Binns: www.binnsindexing.co.uk

Requests to copy any part of this publication should be made to the publisher:
IHS BRE Press
Garston, Watford WD25 9XX
Tel: 01923 664761
Email: brepress@ihs.com

IHS BRE Press makes every effort to ensure the accuracy and quality of information and guidance when it is published. However, we can take no responsibility for the subsequent use of this information, nor for any errors or omissions that it may contain.

EP 99

© Mark Key 2009

First published 2009

ISBN 978-1-84806-107-1

CONTENTS

LIST OF ILLUSTRATIONS		vii
LIST OF TABLES		ix
PREFACE		xi
ACKNOWLEDGEMENTS		xii
ACRONYMS AND ABBREVIATIONS		xiii
CHEMICAL FORMULAE		xiv
GLOSSARY OF TERMS		xv
1	**INTRODUCTION**	1
1.1	Background	1
2	**MASONRY CONSTRUCTION: A SHORT HISTORY**	5
2.1	Introduction	5
2.2	Past to present: the last 150 years	8
2.3	What is sustainable masonry?	13
2.4	Requirements for the future: quality or quantity?	14
2.5	Summary	17
3	**NEW MASONRY COMPONENTS**	19
3.1	Introduction	19
3.2	Embodied energy	20
3.3	Mortar	21
	3.3.1 Production of hydraulic lime	24
	3.3.2 Production of cement	27
	3.3.3 Magnesium oxide cements	31
	3.3.4 Aggregates	31
	3.3.5 Factory-produced mixes	32
3.4	Masonry units	33
	3.4.1 Functionality	34
	3.4.2 Clay bricks	34
	3.4.3 Calcium silicate bricks	39
	3.4.4 Concrete bricks/blocks	41
	3.4.5 Stone	45
3.5	Insulation	51
3.6	Discussion and summary	54

4 DESIGN CONSIDERATIONS — 57
4.1 Introduction — 57
4.2 Design for deconstruction and reuse — 57
4.3 Design for longevity — 64
    4.3.1 Durability and adaptability — 64
    4.3.2 Consideration of whole life costs — 72
    4.3.3 Masonry quality and craftsmanship: do they still exist? — 73
    4.3.4 The quest for enduring aesthetic appeal: the new vernacular — 79
    4.3.5 Design for climate change — 81
4.4 Discussion and summary — 84

5 THE INDUSTRY'S VIEWS — 87
5.1 Introduction — 87
5.2 Interview data — 90
5.3 Summary — 115

6 SUSTAINABLE MASONRY: A DIAGNOSIS — 117
6.1 Introduction — 117
6.2 Quality or quantity? — 117
6.3 New masonry components — 118
6.4 Design considerations — 119
    6.4.1 Design for deconstruction — 119
    6.4.2 Design for longevity and the consideration of whole life costs — 120
    6.4.3 Masonry quality and craftsmanship: do they still exist? — 120
    6.4.4 Enduring aesthetic appeal — 121
    6.4.5 Design for climate change — 122

7 LOOKING FORWARD — 123
7.1 Introduction — 123
7.2 Future needs — 125
    7.2.1 Energy consumption during manufacture — 125
    7.2.2 Reclaimability — 126
    7.2.3 Regulation — 126
    7.2.4 Quality and craftsmanship — 126
    7.2.5 Evolution — 126

REFERENCES — 127

USEFUL CONTACTS — 133

INDEX — 135

LIST OF ILLUSTRATIONS

Illustrations have been provided by the author unless indicated otherwise.

Figure 1: The Court, Queens' College, University of Cambridge — 7

Figure 2: Clay grinding mill manufactured by C Whittaker & Co Ltd of Accrington (extract from original manufacturer's catalogue provided by Richard Collinge, Furness Brick & Tile) — 9

Figure 3: Brick-making machine of 1863, manufactured by Clayton's of Atlas Works, Dorset Square, London (from *Bricks and brick making*, M Hammond, 1990) — 10

Figure 4: Wire cutting machine manufactured by John Whitehead & Co (circa 1879) (from *Brick: a world history*, J W P Campbell and W Pryce, 2003) — 10

Figure 5: Solid wall construction in English bond (from *Building construction*, volume 1, W B McKay, 1938) — 11

Figure 6: Cavity wall construction (from *Building construction*, volume 2, W B McKay, 1944) — 11

Figure 7: Solid wall construction (from *Factsheet 5 – the use of aircrete's solid wall construction*, Aircrete Bureau, 2005) — 12

Figure 8: German manufacturer Ziegel's cellular clay solid wall building system (photo: Rob Woodhouse, Ziegel UK) — 12

Figure 9: Graph showing lower indoor temperatures due to thermally heavyweight construction (from an Arup report, *UK housing and climate change: heavyweight vs. lightweight construction*, C Twinn and J Hacker for Bill Dunster Architects, 2005) — 16

Figure 10: The blaze at Beaufort Park, North London (photo: Rex Features) — 17

The Gillespie Centre, Clare College, University of Cambridge (photo: the Master and Fellows of Clare College, University of Cambridge) — 18

Figure 11: The lime cycle — 24

Figure 12: Chemical process in the production of natural hydraulic lime (from the St Astier Limes website, www.stastier.co.uk, 2001) — 27

Figure 13: A pre-calciner tower, raw meal silo and exhaust stack — 29

Figure 14: Typical cement production using a pre-calciner dry kiln (from *Integrated pollution prevention and control (IPPC) – guidance for the cement and lime sector*, Environment Agency, 2001) — 30

Figure 15: Mortar silos — 33

Figure 16: Samples of clay brick colours and textures (images: Ibstock Brick Ltd) — 35

Figure 17: The Aqua Claudia, Rome — 36

Figure 18: Waste waiting to be crushed to 'grog' — 37

Figure 19: Freshly pressed clay bricks — 37

Figure 20: Traditional brick kiln — 37

Figure 21: Recently fired clay bricks — 37

Figure 22: Ibstock's unfired Ecoterre™ earth brick/block (image: Ibstock Brick Ltd) — 39

Figure 23: Autoclave used to manufacture calcium silicate bricks (© Esk Building Products Ltd) — 40

Figure 24: Side view of autoclave (© Esk Building Products Ltd) — 40

Figure 25: Uncured calcium silicate bricks (© Esk Building Products Ltd) — 40

Figure 26: Samples of calcium silicate brick colours and textures (© Esk Building Products Ltd) — 41

Figure 27: Dense concrete blocks (image: Thomas Armstrong (Concrete Blocks) Ltd) — 42

Figure 28: Lightweight concrete blocks (image: Thomas Armstrong (Concrete Blocks) Ltd) — 42

Figure 29: Aircrete blocks (image: Thomas Armstrong (Concrete Blocks) Ltd) — 42

Figure 30: Silos containing the constituents of aircrete blocks — 43

Figure 31: Aircrete slurry mixer — 43

Figure 32: Slurry setting in steel mould — 43

Figure 33: 'Cake' being cut into blocks by tensioned wires	44	*Figure 55:* A cross-section through the Traditional Plus system (from *Traditional Plus – single and two storey design guide*, CERAM, 2004)	67
Figure 34: Sliced cake ready to be wheeled into an autoclave	44		
Figure 35: Front of autoclaves used to cure aircrete blocks	44	*Figure 56:* Two courses of the composite masonry units developed by Oxford Brookes University and sponsored by Hanson plc (from 'The development of a composite masonry unit for the UK construction industry', in *Masonry International*, Baiche et al, 2007)	68
Figure 36: Cured aircrete blocks	44		
Figure 37: Aircrete blocks being cracked apart and shrink wrapped	44		
Figure 38: Packs of Thomas Armstrong airtec blocks awaiting transportation from the Catterick Plant	44	*Figure 57:* Old pit building on Winscales Moor prior to works starting	69
Figure 39: Drilling of Red St Bees sandstone	47	*Figure 58:* Pit building being given new life as a dwelling	69
Figure 40: Explosive charges being prepared for insertion into holes in the sandstone	47	*Figure 59:* Brand's 'Shearing layers of change' (diagram by D Ryan, from *How buildings learn* by S Brand, © S Brand, 1994. By permission of Weidenfeld and Nicholson, an imprint of The Orion Publishing Group, London, and Viking Penguin, a division of Pengin Group (USA) Inc.)	71
Figure 41: Split sandstone following detonation of explosives	47		
Figure 42: Stone being dragged from base of quarry	48		
Figure 43: Numbered stone awaiting transportation from the quarry	48	*Figure 60:* The rebuilt west elevation of St Pancras Station, London (photo: Irvine-Whitlock)	74
Figure 44: Profiling saw at York Minster stonemasons' yard	48	*Figure 61:* The Jerwood Centre, Grasmere, Cumbria	74
Figure 45: Pneumatic hand tooling of York stone	49	*Figure 62:* Shelter over a house being constructed at Mariners Way, Whitehaven	75
Figure 46: Lanercost Priory, Cumbria	51	*Figure 63:* The finished product – one of the houses built by High Grange Developments	75
Figure 47: Stages of technological improvement; 28-day compressive strength of Portland cement mortar (adapted from *Lea's chemistry of cement and concrete*, P C Hewlett (ed), 2001, with permission from Elsevier)	59		
		Figure 64: English bond	79
		Figure 65: Flemish bond	79
		Figure 66: Stretcher bond	79
Figure 48: Nineteenth century picture of Cockermouth Fire Service (photo: Cockermouth Museum Group)	62	*Figure 67:* Cavity wall built with sandstone facing block external leaf	80
Figure 49: Mitchell's cattle auction building being demolished in 2001 (photo: Graham Edward, Chetwoods, Leeds)	62	*Figure 68:* Extension built with rendered cavity blockwork	80
		Figure 69: Wall insulated on the inside (from *BRE building elements: walls, windows and doors*, H W Harrison and R C de Vekey, 1998)	82
Figure 50: New Sainsbury's supermarket in Cockermouth	62		
Figure 51: Planning and resource efficiency model for demolition and new build (adapted from *A report on the demolition protocol*, EnviroCentre Ltd, 2002)	63	*Figure 70:* The BedZED development, Sutton (photo: Tom Chance, BioRegional)	83
		Figure 71: Passive solar systems (from *Ecohouse 2: a design guide*, S Roaf et al, © Elsevier, 2004)	84
Figure 52: The Circus, Bath (photo: Nick Clarke)	65	Door head detail to restoration and extension of a historic farm complex, Cotswolds (architect: Robert Adam of Robert Adam Architects; contractor: Alfred Groves & Sons)	86
Figure 53: Horizontal cracking of bed joints due to wall tie failure	65		
Figure 54: Corroded wall ties (from *BRE building elements: walls, windows and doors*, H W Harrison and R C de Vekey, 1998)	65		
		De Grey Court, York St John University (photo: Sarah Blee)	116

LIST OF TABLES

Table 1: CO_2 emissions [in 1998] by construction product and materials sector (adapted from Table 8.3.1 in *The construction industry mass balance: resource use, wastes and emissions*, Viridis, 2002) — 2

Table 2: Typical levels of CO_2 embodied in masonry materials (adapted from *Inventory of carbon and energy (ICE)*, G Hammond and C Jones, 2006) — 21

Table 3: Requirements for mortar (from British Standard BS 5628-1, 2005a, © BSI. Reproduced with permission. British Standards are available from www.bsigroup.com) — 22

Table 4: Natural hydraulic lime mortars for use with masonry (from *The use of lime-based mortars in new build*, NHBC Foundation, 2008) — 23

Table 5: Classification of hydraulic limes (from *The English Heritage directory of building limes*, English Heritage, 1997, www.donhead.com) — 25

Table 6: Typical energy consumption of different kiln systems (from *Integrated pollution prevention and control (IPPC) – guidance for the cement and lime sector*, Environment Agency, 2001) — 29

Table 7: Environmentally friendly moves made by clay brick manufacturers — 38

Table 8: Properties of insulation materials — 53

Table 9: Single-leaf masonry: recommended thickness of masonry for different types of construction and categories of exposure (from British Standard BS 5628-3, 2001a, © BSI. Reproduced with permission. British Standards are available from www.bsigroup.com) — 70

Table 10: Interviewees — 89

Stanbrook Abbey, Yorkshire
Designed by 2008 Stirling prize-winner Feilden Clegg Bradley and built for the Benedictine sisters of Stanbrook Abbey. With its local sandstone outer leaf, the new nunnery is located in the North Yorkshire Moors Park, having been taken through the planning process under an 'extraordinary building for extraordinary people clause'

PREFACE

Masonry buildings have inherent durability that has been proved over thousands of years. They offer aesthetic appeal, strength, resistance to fire and to the elements, and allow the maintenance of a suitable indoor environment, both acoustically and thermally.

The early part of the twenty-first century has seen a significant move away from traditional approaches to building, with the UK government choosing to promote modern methods of construction (MMC) as a means of addressing the chronic need for new housing. In some cases, MMC is being put forward as a more sustainable solution in terms of an ability to offer improved energy efficiency and the fact that some prefabricated systems use renewable resources. The high levels of energy needed to manufacture the most common materials used to construct masonry buildings are often cited by advocates of MMC as the main reason that masonry is 'unsustainable', along with claims that the sector does not possess the skill base to provide the quantity of buildings currently being demanded.

In response, the masonry sector has formed the Modern Masonry Alliance to extol the virtues of masonry construction against newer, unproved prefabricated construction systems. While arguing that the government's case is a flawed and extremely short-sighted response to the problem in hand when measured against the current and probable future requirements for completed buildings, the masonry sector has also been forced to come forward with new ideas to counter claims that masonry is an outmoded form of construction. This book examines the ways in which the masonry sector can demonstrate that it has the capability to produce sustainable buildings that not only match the environmental targets set by the government, but also go beyond them, and by doing so create masonry buildings that have the lowest possible impact on the environment, both now and for many future generations. The book considers the following four major areas:

1. Are manufacturers of materials used for masonry ensuring that their production methods have the minimum possible impact on the environment, and can they do more in this respect?
2. What can the designers/specifiers of masonry buildings do to ensure that the masonry has the minimum possible impact on the environment, both now and for future generations, while also ensuring that Building Regulations can be met by both the initial build and future refurbishment?
3. Are common methods of masonry construction tangible options for the future when taking into account our changing climate and the quest for optimum levels of sustainability? If not, what are the alternatives?
4. Does government guidance and planning legislation do enough to encourage the construction of sustainable masonry buildings in terms of the effect their initial construction has on the environment? In light of these buildings' ability to achieve a significant lifespan, is the achievement of enduring aesthetic appeal in terms of vernacular and style adequately encouraged?

If the masonry sector now pauses to consider new ways to meet these four objectives, instead of following a path by which the increasing requirement for insulation dictates the form of masonry structures, it has the capability to ensure that masonry maintains its position as the most common and widely respected form of construction known to man.

ACKNOWLEDGEMENTS

I would like to thank my wife, Shirley, and children, Daniel and Holly, for their love and support. I would also like to thank the following organisations and people for their kind assistance in the preparation of this book:

Ian Abley, audacity
Michelle Beech, Thomas Armstrong Ltd
Dr Neil Beningfield, Neil Beningfield & Associates Ltd
Michael Burdett, York College
Kevin Calpin, York College
Andy Clark, Thomas Armstrong (Concrete Blocks) Ltd
Martin Clarke, British Precast
Nick Clarke, IHS BRE Press
Richard Collinge, Furness Brick & Tile Co Ltd
Copeland Borough Council (the author's employer)
Paul Dewick, Irvine-Whitlock
Michael Driver, Brick Development Association
Jim Dunn, Holloway White Allom
Professor Geoff Edgell, CERAM
Graham Edward, Chetwoods
Gloria Edwards, Cockermouth Museum Group
Kelly Frost, Stone Federation Great Britain
Cliff Fudge, H+H Celcon
Dr Jacqui Glass, Loughborough University
Gemma Hall, Allford Hall Monaghan Morris
Dr David Hills, Ibstock Brick Ltd

Stafford Holmes, Rodney Melville & Partners
John Houston, BRE
Nicholas Kehagias, Ash Sakula Architects
Marion Kerfoot, IHS BRE Press
Paul Livesey, Castle Cement
Dr Sebastian Macmillan, University of Cambridge
Paul Monroe, York College
Mark Oliver, H+H Celcon
David Peacock, Bluestone
Ian Pritchett, Lime Technology
Dr Michael Ramage, University of Cambridge
Professor Sue Roaf, Oxford Brookes University
Kathryn Sanderson, Rivington Street Studio Architecture
Charles Thomson, Rivington Street Studio Architecture
Brian Timperley, Manchester Testing Laboratories
Matthew Todd, Copeland Homes
Murray Treece, Esk Building Products Ltd
Dr Toby Wilkinson, University of Cambridge
Rob Woodhouse, Ziegel UK Ltd

ACRONYMS AND ABBREVIATIONS

BCA	British Cement Association
BDA	Brick Development Association
BedZED	Beddington Zero Energy Development
BGS	British Geological Survey
BRE	Building Research Establishment
BSI	British Standards Institution
BURA	British Urban Regeneration Association
CERAM	abbreviation now used by British Ceramic Research Ltd
CIOB	Chartered Institute of Building
CIRIA	Construction Industry Research and Information Association
DCLG	Department for Communities and Local Government
DETR	Department of the Environment, Transport and the Regions
GLA	Greater London Authority
HL	artificial hydraulic limes
HLM	hydraulic lime mortar
MMA	Modern Masonry Alliance
MMC	modern methods of construction
NHL	natural hydraulic limes
NHL-Z	special natural hydraulic limes
NVQ	national vocational qualification
ODPM	Office of the Deputy Prime Minister
PFA	pulverised fuel ash
PVC	polyvinyl chloride
R&D	research and development
WLC	whole life cost(s)
WRAP	Waste and Resources Action Programme

CHEMICAL FORMULAE

Al_2O_3	alumina
C_2S	dicalcium silicate
C_3A	tricalcium aluminate
C_3S	tricalcium silicate
C_4AF	tetracalcium aluminoferrite
$CaCO_3$	calcium carbonate (limestones, chalks, shells and corals)
CaO	calcium oxide (quicklime)
$Ca(OH)_2$	calcium hydroxide (slaked lime, hydrated lime or lime putty)
CO_2	carbon dioxide
Fe_2O_3	iron oxide
H_2O	water
SiO_2	silica

GLOSSARY OF TERMS

Aggregate	the hard filler material, such as sand, in mortars.
Argillaceous	containing clay minerals.
Ashlar	rectangular blocks of stone sculpted to have square edges and even surface appearance. Finely worked and jointed to given dimensions and laid in courses.
Batch	the constituents making up a single load for mixer.
Batching	the process of adding the constituent batch materials to the mixer in the prescribed quantities.
Binder	the material that forms the matrix between aggregate particles in a mortar. A fluid paste when first prepared, it must harden to hold the aggregate in a coherent state.
Calcareous	containing or resembling calcium carbonate; chalky.
Calcine/calcination	to heat a substance so that it is oxidised, reduced or loses water. (The term is used in this book to describe the conversion of calcium carbonate to lime.)
Calcium lime	also termed 'air lime'. Hydrated lime (in powdered form) or lime putty that sets through carbonation rather than a chemical reaction with water.
Carbonation/carbonated	the process of forming carbonates. A calcium lime mortar is said to have carbonated when the binder has reacted with carbon dioxide from the air to form calcium carbonate and developed strength beyond that achieved simply by drying out.
Cementitious	having cementing properties (ie will set and harden in the presence of water).
Curing	the hardening process of a plastic mortar mix containing a cementitious binder.
English bond	the arrangement of masonry units in which alternate courses of headers and stretchers are visible in a wall.
Flemish bond	the arrangement of masonry units in which, in each course, alternate headers and stretchers are visible in a wall.
Free lime	calcium lime in a mortar that has not yet carbonated or combined with a pozzolan.
Hydraulic	an ability to set under water.
Quicklime	lime that has been burned but not slaked. It is termed 'quick' because it reacts violently when mixed with water; calcium oxide.
Pozzolan	a material containing constituents, generally reactive aluminates and silicates, that will combine with lime at normal temperatures in the presence of moisture to form stable binding hydrates.
Random rubble wall	a wall of rubble stone laid uncoursed.
Siliceous	material containing abundant silica.
Slake/slaking	the action of combining quicklime with water.
Stretcher bond	the arrangement of masonry units in which only stretchers are visible in a wall.
Thermal mass	the capacity of a material to store heat in response to the surrounding environmental conditions.

The Hothouse, Hackney
An Arts Trust building designed by Ash Sakula Architects. This highly regarded project has won a Civic Trust Award and was shortlisted in the 'best commercial building' category of the Brick Awards 2009

1 INTRODUCTION

Perhaps the most powerful statistic is that provided by assessments of the environmental footprint of mankind's activities – the area of land that would be needed to provide all our materials and energy requirements, and to deal with the disposal of waste. London, for example, would need an area of land about the size of Spain to be fully sustainable at present levels of consumption. If every country were to have the same environmental impact as Western countries currently do, we would already need more than three Earths to ensure our long term, sustainable survival. Clearly mankind has to do something.

Addis (2006, p. 5)

1.1 BACKGROUND

Half of all resources consumed across the planet are used in construction, making it the least sustainable industry in the world (Edwards, 2001). In the UK alone, at least 70 million tonnes of construction and demolition waste is created each year, representing approximately 17% of total waste production (CIRIA, 2001). Among this waste are vast amounts of potentially reusable masonry components. The design and layout of buildings such as supermarkets can become obsolete after a relatively short period, although the masonry units used in their construction are likely to have a lifespan far in excess of the useful life of such buildings. It is clear that something has to change, and that clients, contractors and building designers have an important role to play in such change.

The government, in recognising the wasteful performance of the UK construction industry, has introduced many initiatives to tackle the problem, including markedly increasing the amount of landfill tax payable per tonne.

The production of materials used for masonry also has a massive effect on the environment. The embodied energy contained in around 10 old bricks is equivalent to that contained in a gallon of petrol, and although 3.5 billion bricks are manufactured each year in the UK, 2.5 billion are destroyed (Woolley et al, 1997). Of these, only about 140 million are salvaged and reused; the remainder are consigned to landfill (Addis, 2006). To put this into perspective, the energy lost in the bricks going to landfill each year would be enough to run around 850 000 cars for one year in the UK – equivalent to around 250 million gallons of petrol (IAM Motoring Trust (online)). Global warming is thought to be taking place as a result of burning fossil fuels in processes that release carbon dioxide (CO_2) and other greenhouse gases into the atmosphere. Worldwide, it is estimated that 5.4% of all CO_2 emissions result from cement production – around 1.4 billion tonnes every year (Pritchett and Dukes, 2003). Numerous targets have been introduced to reduce CO_2 emissions in the UK, and overall the government has set a target of a 60% reduction in CO_2 levels by the year 2050 in its climate change bill. This bill will eventually give the UK the world's first legal framework for transition to a low-carbon economy (BBC News (online) (a)). As Table 1 shows, the cement industry easily tops the UK construction product and material sector's league of highest CO_2 producers, with other masonry-related component activities also scoring highly.

It may be said that traditional masonry has had its day, facing as it does stiff competition from modern methods of construction (MMC). However, research suggests that the vast majority of UK house buyers prefer homes built with masonry, believing them to be more robust and 'permanent' than those constructed with prefabricated components (MORI, 2001). In addition to this survey, a study by Arup Research & Development (2005) found that heavyweight constructions such as masonry will be required in future years to combat global warming, their thermal mass being ideal in warm climates due to their ability to absorb and store heat. However, the vast amounts of raw materials and energy needed to produce components, along with the copious amounts of CO_2 emitted into the atmosphere during production, mean that masonry buildings in their present guise leave room for improvement in terms of a sustainable approach.

Table 1: CO_2 emissions [in 1998] by construction product and materials sector (kilotonnes per annum)

Reference/activity	CO_2
Cement products	12 464
Finishes, coatings, adhesives	2329
Quarry products/mineral extraction	1513
Plastic products	1058
Wood products	918
Fabricated metals	825
Ceramic products	751
Clay-based products	690
Glass-based products	394
Cabling, wiring, lighting	93
Stone and other non-metallic mineral products	52

The need to examine more sustainable options for masonry construction is not in doubt. Industry cannot continue to extract non-renewable materials as if supplies were endless when this is patently not the case. Neither can it continue to spew vast amounts of CO_2 into the atmosphere on an indefinite basis.

This book is derived from a thesis written by the author while participating on the Interdisciplinary Design for the Built Environment (IDBE) Masters programme at the University of Cambridge, and also draws on previous research work into the merits and demerits of lime- and cement-based masonry mortars. The background work for the book involved the examination of many hundreds of relevant texts and consultation with leading experts and organisations over a four-year period. *Sustainable masonry construction* examines ways in which the UK masonry sector can not only help to achieve the overall objectives set by the UK government, but also go beyond them, and by doing so create components and ultimately low- to medium-rise masonry buildings (up to four storeys in height) that have the lowest possible impact on the environment, both now and for many future generations. The predominant use of masonry walling in the UK is in low- to medium-rise buildings and for the cladding of steel and concrete frame structures, which is why the book limits itself to such structures.

Many masonry buildings are designed and built with little regard for their impact on the environment. Thought is rarely given to the amount of energy embodied in the materials used, the use of reclaimed materials, the accurate expected lifespan of a building (including possible future adaptability), orientation and construction to suit present and possible future climatic conditions, and the deconstruction and reuse of components at the end of a building's useful life.

The book considers four major areas:

1. Are manufacturers of materials used for masonry ensuring that their production methods have the minimum possible impact on the environment, and can they do more in this respect?
2. What can the designers/specifiers of masonry buildings do to ensure that the masonry has the minimum possible impact on the environment, both now and for future generations, while also ensuring that Building Regulations can be met by both the initial build and future refurbishment?
3. Are common methods of masonry construction tangible options for the future when taking into account our changing climate and the quest for optimum levels of sustainability? If not, what are the alternatives?
4. Does government guidance and planning legislation do enough to encourage the construction of sustainable masonry buildings in terms of the effect their initial construction has on the environment? In light of these buildings' ability to achieve a significant lifespan, is the achievement of enduring aesthetic appeal in terms of vernacular and style adequately encouraged?

In looking at the four areas highlighted above, the following chapters draw comparisons with current methods of masonry construction, while at the same time taking into account functionality and the possible financial implications of more sustainable options for developers. All relevant factors are balanced against the current and future needs of society and the environment in an attempt to examine the most sustainable but functional methods of masonry construction for low- to medium-rise buildings that will also prove to be economically viable and functional.

The following chapters will offer:

- A short history of masonry construction leading to the current predominant use of cavity wall construction (Chapter 2). This chapter also defines the requirements for sustainable masonry and sets out, from a starting point of today's construction climate, future requirements for more sustainable masonry buildings.
- A critical appraisal of new masonry components (Chapter 3). This chapter considers how much is being done to minimise the environmental impact of manufacturing processes and what more could be done by manufacturers in this respect.
- An appraisal of design options (Chapter 4). This chapter examines how materials with a low environmental impact, including those reclaimed from demolished buildings, can be incorporated into building design options that make the best of such materials by way of ensuring that the completed buildings have the minimum environmental impact and consider the eventual reuse of the materials used in their construction.
- The views of leading industry experts obtained via interviews undertaken by the author (Chapter 5).
- An overview of the material covered in Chapters 1–5 (Chapter 6).
- A look at how the masonry sector of the future might achieve the objective of 'sustainable masonry construction' (Chapter 7).

2 MASONRY CONSTRUCTION: A SHORT HISTORY

Buildings are our third skin. To survive we need shelter from the elements using three skins. The first is provided by our own skin, the second by a layer of clothes and the third is the building. In some climates it is only with all three skins that we can provide sufficient shelter to survive, in others the first skin is enough. The more extreme the climate, the more we have to rely on the building to protect us from the elements. Just as we take off and put on clothes as the weather and the climate changes so we can alter our buildings to adapt to changes in climate.

Roaf (2004, p. 13)

2.1 INTRODUCTION

Masonry construction is one of man's oldest construction methods. Most early applications used natural stone assembled in a 'dry stack' process. Crude mortars were developed by civilisations such as the Egyptians and Persians, using gypsum and lime as binders with sand and water to create a bond between stone masonry units (Mortar Industry Association, 2004a). There is firm evidence of the use of lime dating from about 6000 BC: a floor excavated in the 1960s in Lepenski Vir in the former Yugoslavia, which consisted of a type of mortar made from lime, sand, clay and water (Oates, 1998). It is well recognised that lime was the principal binder for mortars in the UK for nearly 2000 years, the Romans having introduced the technology following their invasion of the British Isles in 43 AD. The earliest known explanation of the reactions by which limestone becomes cementitious on burning is given by the Roman engineer Vitruvius in the first century AD (in Bogue, 1947, p. 5):

Stones, as well as all other substances, are compounded of the elements; those which have most air are tender; those which have most water are, by reason of their humidity, tenacious; those which have most fire, brittle. If these stones were only pounded into minute pieces, and mixed with sand without being burnt, they would not indurate or unite; but when they are cast in the furnace, and these penetrated by the violent heat of the fire, they lose their former solidity; being calcined and deprived of their strength they are left exhausted and full of pores. The water and air, therefore, which are in the substance of the stones, being thus discharged and expelled, and the latent heat only remaining, upon being replenished with water, which repels the fire, they recover their vigor and the water entering the vacuities occasions a fermentation; the substance of the lime is thus refrigerated and the superabundant heat ejected.

The Industrial Revolution in the eighteenth century heralded the beginning of the end of the extensive use of lime mortars. Prior to this time, the vast majority of lime mortars had relied upon carbonation (the absorption of CO_2 from the air) in order to set. For a hydraulic set, the Romans ground together lime and volcanic ash or finely ground burnt clay tiles. The active silica and alumina in the ash and the tiles combined with the lime to produce what became known as pozzolanic cement, from the name of the village of Pozzuoli, near Vesuvius, where the volcanic ash is found. The complex engineering projects proposed at the time of the Industrial Revolution required mortars with hydraulic properties. This led to the first hydraulic lime products, developed by Joseph Smeaton in 1756 by calcining Blue Lias limestone containing clay; the argillaceous and/or siliceous material activated within the clay during burning of the limestone giving these new mortars their hydraulic reactivity (Hewlett, 2001). These mortars were used by Smeaton to construct Eddystone Lighthouse off the Cornish coast between 1756 and 1759, the structure standing for 127 years until it was dismantled stone by stone and reassembled at Plymouth Hoe (Mortar Industry Association, 2004a).

Numerous refinements of these initial hydraulic mixes were made during the ensuing 68 years until 21 October 1824 when Joseph Aspdin, a bricklayer from Leeds, took out British Patent 5022 for the first Portland cement, which became the basis for many of today's modern masonry mortars, although lime would continue to be used as the principal binder for mortars until the early part of the twentieth century (Hewlett, 2001).

Stone has been used to construct buildings since the beginning of the history of civilisation, but the desire for uniformity in the size and shape of masonry units and the wide availability of clay led to the development of brick. Brick was first fabricated from mud and baked in the sun between 10 000 and 8000 BC, and the first clay bricks fired in a kiln were manufactured in Mesopotamia around 3000 BC (Campbell and Pryce, 2003). Brickwork was used extensively in Roman times, but it was not until the end of the thirteenth century that brickwork in England began to be used on a limited scale. The Tudor and Jacobean periods then saw a massive increase in its use.

The Great Fire of London in 1666 was the main factor that influenced the extensive development of brickwork in the second half of the seventeenth century in England (Briggs, 1925), and from that point onwards, clay bricks became the predominant masonry unit used up until the middle of the twentieth century (Harrison and

Figure 1: The Court, Queens' College, University of Cambridge, which remains one of the finest examples of medieval brickwork in the UK, having been virtually untouched since its completion in 1448

de Vekey, 1998). More recently, calcium silicate bricks (made from hydrated lime and sand) and concrete masonry bricks and blocks (made from cement, sand, gravel or crushed aggregate, water and a series of admixtures) have been developed and used.

The many age-old masonry buildings in existence throughout the UK bear testament to the durability and appeal of this tried-and-tested method of construction (Figure 1).

Masonry units are bonded together in order to form a monolithic wall without continuous vertical joints. When built, a modern masonry building must have aesthetic appeal, adequate strength, be fire resistant, have adequate resistance to the elements, resist erosion and corrosion for the life of the building, allow the maintenance of a suitable indoor environment, and offer acoustic and thermal properties (Harrison and de Vekey, 1998). To achieve all of these properties, masonry construction has had to evolve markedly since the middle part of the nineteenth century.

2.2 PAST TO PRESENT: THE LAST 150 YEARS

Masonry in the form of brickwork, blockwork and natural stone is one of the most familiar construction materials. Perhaps it is for this reason that it tends to be taken for granted and that there is a feeling that little thought needs to be given to the design and building of masonry walls. This of course is far from the case.

Hendry and Khalaf (2001, Preface)

From the point of view of quality control and production technique, the most important event in the history of masonry mortars was the development of the first rotary kiln by the American Frederick Ransome in 1885 (Blount, 1920). This replaced the shaft kiln, and cement manufacture was changed from a batch process to a continuous production process. This change took place slowly at the beginning of the twentieth century, and in the ensuing 100 years, following the erection of the first rotary kilns in the UK, kiln reaction stages have undergone many changes directed by aspects of quality and economy. The changes have involved a reduction in feedstock water content, the nature of the firing flame, fuel economy and, later, computerised operation, resulting in the normal Portland cements we know today. These cements are quality calcareous cements manufactured in a rotary kiln and can be defined as follows:

Normal Portland cement is a synthetic mixture of calcium silicates formed in a molten matrix from a suitably proportioned and homogeneously prepared mixture of calcareous and argillaceous components.

Hewlett (2001, p. 11)

As the technical knowledge of normal Portland cement increased, special cements for specific applications were devised by relating mineral components to performance-related properties, and the name 'Portland cement' achieved the status of a generic term.

By the mid-1930s, great concern was being expressed in the *Journal of the Royal Institute of British Architects* about the quality of hydraulic lime mortars, the journal of 7 July 1934 stating that:

The breaking down of the old localised traditions, particularly of lime manufacture and use has resulted both in operatives having to use materials with which they are unfamiliar and employing known methods with unsuitable materials. At the present time very many different kinds of lime are produced and distributed over wide areas. There is as yet no standard specification and no organised instruction on methods of use. It is true to say that the question of making and using building lime is today chaotic. Therefore in talking of 'lime mortar' one is generalising dangerously since there is little guarantee that any two limes will give identical results. It is good news that the Building Research Station are making serious efforts to evolve a reasonable order in this matter. It is therefore small wonder that architects attempt to play for safety by specifying Portland cement mortars. Portland cement is a uniform product made to a standard specification and its qualities are known; also it makes for rapid building. Moreover building surveyors encourage its use partly for the same reason and partly because it 'gives strength', a factor with which they are mainly concerned.

In addition to the concerns expressed by the *Journal of the Royal Institute of British Architects*, the Brick Development Association or BDA (2001) suggests that the demise of lime mortars was also hastened in the 1930s by a move away from heavy solid wall construction, to thinner walls permitted by structural engineering developments at that time. These developments required mortars to be stronger in both compression and flexure – cement mortars being an ideal material in these circumstances.

Portland cement, a material manufactured to prescribed standards, provided consistent quality and mortars of more predictable performance – factors which led to its wide acceptance and the decline of traditional lime mortars. Since the Second World War Portland cement mortars have dominated construction practice.

Smith (1999, p. 29)

In 1973, lime mortars were removed from CP 121.101 (*Code of practice for brickwork*) because they were not compatible with the speed of modern construction (BDA, 2001). British Standard BS 5628 (*Code of practice for use of masonry*) replaced CP 121 in 1978 and this code, which is still in use, also does not cover the use of lime mortars. A Draft for Development, *The structural use of unreinforced masonry made with natural hydraulic lime mortars* (NHBC Foundation, 2008), has been written by enthusiasts of hydraulic lime mortars as a technical annex for use with British Standard BS 5628-1 (*Code of practice for use of masonry – Structural use of unreinforced masonry*) (BSI, 2005a).

Prior to the development of a mechanised, factory-based manufacturing process in the nineteenth century, bricks were handmade from clays dug from the ground. As sufficient quantities of suitable clays were accessible throughout the UK, bricks were manufactured in every English county and used locally. As the nineteenth century unfolded, steam power, and later electricity, facilitated the development of machines that could mould, press, grind and extrude clay (Figures 2–4) and kilns that could quickly fire greater quantities of bricks with more consistency. The main mode of transport at that time was the horse and cart, which made the movement of heavyweight materials extremely difficult (Hammett, 1991). However, even following the evolution of improved transportation in the form of narrow boat, rail and then road vehicles, bricks were still being used within a 30-mile radius of where they were made at the time of the Second World War. Following the war, a more rationalised brick-making industry emerged, and from that

Figure 2: Clay grinding mill manufactured by C Whittaker & Co Ltd of Accrington. Installed in 1901, one of the mills is still in use by Furness Brick Ltd in Cumbria (see also Chapter 3, 'Clay bricks')

Figure 3: Brick-making machine of 1863, manufactured by Clayton's of Atlas Works, Dorset Square, London

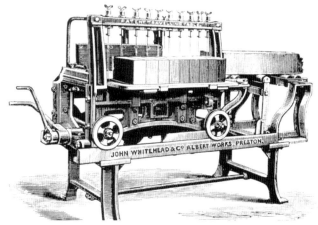

Figure 4: Wire cutting machine manufactured by John Whitehead & Co (circa 1879)

point on, bricks have been made by a small number of manufacturers and marketed and transported throughout the UK. Clay remains the dominant material for brick making in the UK today, accounting for nearly 90% of all the bricks used in the UK (Hammett, 1991).

The Second World War not only had a massive effect upon the move away from locally produced materials, but also reshaped the way masonry buildings were constructed. Although the cavity wall originated in the nineteenth century, having been developed to alleviate the problems of water penetration being experienced through some solid wall constructions, it was the mass-housing projects required following the war that accelerated the move towards cavity construction, as opposed to the more traditional solid wall used to build the vast majority of low- to medium-rise masonry buildings up to that time (BDA, 2001). It seems that, for the construction industry in post-war Britain, cavity wall construction and the stronger, more consistent cement-based mortars required for their construction were a match made in heaven. Figures 5 and 6, taken from the McKay *Building construction* series of textbooks published in the 1930s and 1940s (McKay, 1938 and 1944), detail the two typical construction types still being used at that time (solid and cavity wall).

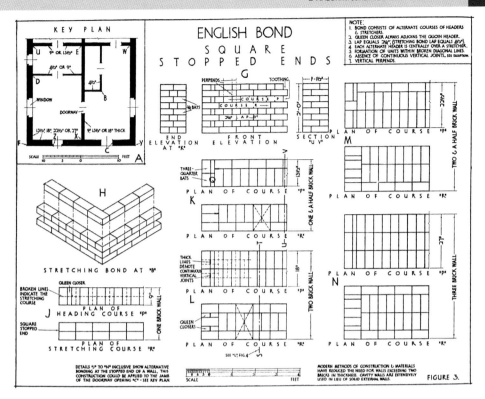

Figure 5: Solid wall construction in English bond

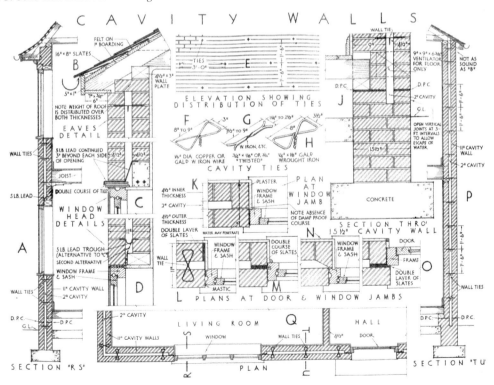

Figure 6: Cavity wall construction

Since the 1980s, there has been an increasing requirement for thermal insulation in buildings, thereby precluding the use of dense brickwork for the internal leaf of cavity walls. Lightweight blocks, whose thermal performance far exceeds that of brickwork, are now used in lieu of brickwork on the internal leaf for new construction. In the decade leading up to the 2006 changes to Part L of the Building Regulations (*Conservation of fuel and power*), the U-value required for external walls improved by one-third, from 0.45 W/m²K to 0.30 W/m²K. As a result, cavities being either partially or fully filled with insulation are, on average, 100 mm wide, as opposed to the 50 mm cavity detailed in Figure 6. Having said this, some cavities can be much wider, depending upon the energy efficiency requirements of the building designer or client, with cavities of up to 300 mm not being uncommon (Lazarus, 2002).

As cavity construction becomes ever more complicated due to the UK government's attempts to move towards a low-carbon economy, there is a growing interest in a return to the more traditional solid wall form of masonry construction, which is still commonly used in many parts of Europe (Smith, 2006). In most cases, however, modern solid wall construction (Figures 7 and 8) bears no resemblance to the solid walls constructed up to the middle of the twentieth century, which tended to be thick masonry made up with brick or coursed/random rubble stone that was either rendered or faced externally.

Figure 7: Solid wall construction using thin-joint construction with lightweight aircrete blocks (right), which can then be covered by tiles (far right) or insulated and rendered externally to reduce the risk of water penetration

Figure 8: German manufacturer Ziegel's cellular clay solid wall building system

2.3 WHAT IS SUSTAINABLE MASONRY?

Sustainable development to which many in Europe aspire accepts the principles of nature, seeks limits to corporate growth and personal consumption, and uses the minimum of resources to achieve the maximum in environmental quality.

Edwards (1999, p. 6)

The drive for more sustainable development is perhaps the defining issue of the early twenty-first century. If masonry is to retain its place as a primary construction method, there will need to be a major shift towards a more sustainable approach, as the vast majority of resources that the manufacturers of masonry components currently use in abundance are not renewable.

Although many have tried, it is extremely difficult to place an absolute definition on the term 'sustainability', so numerous are the issues that surround it. As Spence *et al* (2001, p. 3) state:

Sustainability is an elusive concept, but clearly demands from architects, engineers and constructors a deeper understanding of the interaction of the building with its local and the global environment over a period of time which stretches throughout and even beyond the life of the building; and it calls for a holistic view of the design process on the part of all contributors to it.

Perhaps the most famous and widely used definition of sustainable development is taken from the Brundtland Report (Brundtland, 1987, p. 24):

Development which meets the needs of the present without compromising the ability of future generations to meet their own needs.

The design and construction of new 'sustainable masonry buildings' should meet as many of the following criteria as possible (Regenerate, 2006; Lazarus, 2002 and 2005; Arup Research & Development, 2005; Institution of Structural Engineers, 1999; Edwards, 1999; CIOB, 2002; Woolley *et al*, 1997):

- Use of local materials (to reduce transport requirements) with low embodied energy levels
- Use of reclaimed materials where possible or components containing recycled material
- Design for long life in order to minimise maintenance requirements and the quantity of materials needed to replace defective components
- Design to create a comfortable and pleasant building for occupants
- Design to ensure easy refurbishment and extension when required in order to prolong the useful life of a building

- Design to minimise operational energy consumption, including the physical orientation of buildings in order to maximise their capacity to exploit renewable energy
- Design for deconstruction and the eventual reuse of as many of the building's components as possible

However, in most cases, the most sustainable option of all is the refurbishment and reuse of existing masonry buildings as the energy required to demolish, manufacture new materials and rebuild is saved. This could essentially be termed the 'recycling of a building' (Roaf *et al*, 2004; Highfield, 2000; Edwards, 1999).

But do the above requirements for a sustainable masonry building fit in with the aims of the UK government and the requirements of society in general?

2.4 REQUIREMENTS FOR THE FUTURE: QUALITY OR QUANTITY?

Last December, Ruth Kelly announced a target for all new-build dwellings to be carbon neutral within 10 years. I would like to raise the bar of that challenge – to make all new-build dwellings carbon neutral from the first stage of construction to the last stage of demolition, with a target life of at least 200 years and to be able to sustain a 4°C rise in summer temperatures over that period. Without some strong intervention from government, we may be constructing buildings that will not only be difficult to maintain, but that will be barely habitable in the hotter summer months that we are almost certainly going to see.[*]

Jones (2007, pp. 34–35)

As detailed in Chapter 1, the UK government announced plans for its climate change bill in an attempt to stem the move towards global warming. This announcement followed quickly in the footsteps of *The economics of climate change: the Stern Review* (Stern, 2007), which gained worldwide attention upon release, primarily (and perhaps sadly) because of the projected detrimental effect that climate change might have on the world's economy, rather than its pure environmental effects. Stern predicts that if no action is taken to reduce greenhouse gas emissions, a global average temperature rise of over 2°C would result by the year 2035, with a 50% chance that the temperature rise would exceed 5°C in the longer term.

---

[*] On 13 December 2006, Ruth Kelly (Secretary of State for Communities and Local Government, May 2006–June 2007) announced in the consultation document *Building a greener future: towards zero carbon development* that building regulation would be tightened over a 10-year period to meet the target of carbon-neutral dwellings by 2016.

Many concerns have been expressed that the UK government is advocating the wrong approach to the construction of new buildings, especially as it has become widely recognised that the climate is already changing, the UK having recently experienced its 10 warmest recorded years, between 1995 and 2006 (BBC News (online) (b) and BBC Weather (online)). Despite this, the government has, in an attempt to push developers to produce housing cheaply and quickly to meet rising demand (especially prevalent in the South-East of England), been championing the use of lightweight prefabricated system building or MMC, which have been defined as:

... those which provide an efficient product management process to provide more products of better quality in less time.

BURA (2005, p. 9)

Unfortunately, although such construction methods do lend themselves to quicker construction times, they have been proved to perform poorly in warm conditions, are more expensive than traditional methods and are of unknown durability (National Audit Office, 2005; BRE, 2005).

A study by Arup Research & Development (2005) found that the types of buildings now being constructed in the UK (and particularly in the South-East) are not well suited either to the climatic conditions being experienced now or to those projected for the future. The report suggests that if a massive requirement for future mechanical cooling of buildings and a subsequent associated rise in energy use is to be avoided, buildings with high levels of thermal mass should be constructed. The general recommendations of the report include:

- Large areas of rooms exposed to heavyweight thermal mass to absorb excess heat gains
- Generous volumes of cooler air ventilation, particularly during night time with the increasing diurnal temperature swing
- Good opaque shading to block direct solar gain, ideally outside the occupied rooms
- Ventilation limited during the hot periods of the day
- Bedrooms positioned north, facing away from direct solar gain

Figure 9 shows how significantly lower indoor temperatures can be achieved during warm summer months by utilising heavyweight construction methods such as masonry.

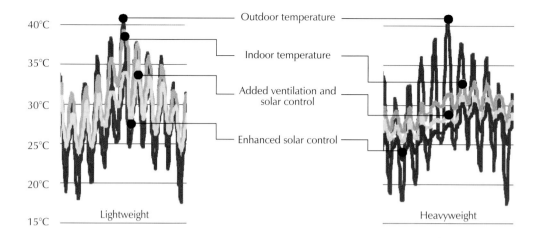

Figure 9: Graph showing lower indoor temperatures due to thermally heavyweight construction

> *Our climate is already changing. Our planet is already warming up. Homes account for over a quarter of our carbon emissions – that is why it is so important that new homes meet much higher standards in future. We need a revolution in the way we build new homes – both to cut carbon emissions and to respond to our changing climate ... Last summer there were reports of big increases in people buying portable fans and air-conditioning units, just to keep cool. The homes of the future need to be designed for hot summers as well as cold winters. We should be building green houses not greenhouses for future generations.*
>
> Yvette Cooper MP (part of the 'Zero-carbon homes – delivering the agenda' speech made on 9 January 2007 while in office as Minister for Housing; DCLG, 2007 (online))

> *... what is the typical architectural response to the challenge of global warming? It is not to make the building do more of the work in providing better shelter against climate change, nor to use solar technologies, but to install air conditioning, which is a key element in the vicious circle that is creating global warming.*
>
> Roaf et al (2004, p. 8)

As well as question marks against the cost, durability and performance of MMC, insurers have expressed concerns over the resilience of prefabricated systems to typical threats such as fire, burst pipes, wind, flooding and theft. Mortgage lenders also have specific concerns about the durability and whole life cost (WLC) of maintaining such buildings and the effect that this may have on their value (BRE, 2005).

Figure 10: The blaze at Beaufort Park, North London

Concerns over fire resistance were found to be justified when, on 12 July 2006, a major fire requiring the attendance of 20 fire service pumps occurred at two timber-framed buildings being constructed at the Beaufort Park development in Colindale, North London (Lane, 2006). The blaze not only destroyed the two buildings but spread to a student halls of residence, damaged Hendon Police College on the other side of the road, and resulted in 30 cars being written off and 2500 people being evacuated (Figure 10).

2.5 SUMMARY

Masonry buildings have high levels of durability that have been proved over many thousands of years. They offer aesthetic appeal, strength, fire resistance, resistance to the elements and allow the maintenance of a suitable indoor environment, both acoustically and thermally.

Although masonry construction has evolved markedly since the middle of the nineteenth century to meet the changing requirements of society and legislation, the time has come to consider how twenty-first century masonry buildings can best meet the needs of the environment, as well as those of present and future generations of society. While the government's recent promotion of MMC might be seen as a short-term answer to a housing shortage, it is likely that MMC's long-term legacy will be a drain on resources in the form of energy needed to cool lightweight buildings and the labour and new materials required to replace them following a relatively short lifespan.

The Gillespie Centre, Clare College, University of Cambridge
Designed by van Heyningen and Haward Architects, the building has won a 2009 RIBA Award and was shortlisted in the 'sustainability' and 'best educational building' categories of the Brick Awards 2009

3 NEW MASONRY COMPONENTS

Once the design strategy is clear, local building materials and those requiring the minimum processing should be selected in preference to highly processed materials and those from further afield ... The durability of materials is also very significant as it will affect the lifespan of a building and, the longer a low-energy house lasts, the less relative impact its materials will have ... Ideally, all building materials should be easily recyclable.

Roaf *et al* (2004, p. 64)

3.1 INTRODUCTION

Construction materials make up over half of the UK's resource use by weight and account for 30% of all road freight. The environmental impacts of extraction, processing and transportation are major contributors to greenhouse gas emissions and resource depletion (Lazarus, 2002).

The most important characteristics that should be sought from new masonry materials to achieve low environmental impact are the ability to facilitate eventual reuse, the use of recycled ingredients, standardisation in order to reduce cutting and subsequent wastage, durability and minimum levels of embodied energy (Regenerate, 2006; Lazarus, 2002 and 2005; Institution of Structural Engineers, 1999; CIOB, 2002; Oxley, 2003).

Ultimately, the best way to ensure that masonry construction has the lowest possible impact on the environment is to reuse the energy previously invested in materials by way of adapting and reusing an existing building or, if this is not possible, reusing materials reclaimed from the demolition of a building that has come to the end of its useful life. The reason for not considering reclaimed materials among possible masonry components for new build in this chapter is that the opportunity to reclaim some of the materials available today may, it is sad to say, have already been lost. This is due to the lack of 'as-built' information available on the buildings currently being demolished and the fact that, due to this lack of knowledge, problems exist in obtaining warranties for reclaimed materials with unknown characteristics (Addis, 2006; CIRIA, 1999a; Woolley *et al*, 1997). Another reason for the current preference for new materials is summed up by Hobbs and Collins (1997, p. 1):

In recent years there has been considerable interest and research into reuse and recycling of materials during the demolition and construction of buildings. However, little progress has been made in the practical implementation of research and government recommendations.

Therefore, the current options for the use of reclaimed materials will be considered under 'Design for deconstruction and reuse' in Chapter 4.

This chapter considers the effects that the production and use of new masonry materials have on the environment and how these may be reduced, starting with the important but sometimes overlooked subject of embodied energy.

3.2   EMBODIED ENERGY

As a global average, it is estimated that 50–60% of the world's annual energy production is used by the industries of the built environment of which 36% is consumed directly in the operation of buildings. A simple calculation then tells us that between 14% and 24% is embodied in the buildings and materials themselves. It is likely that these figures understate the full impact as transportation energy is measured within a different boundary.

Best and de Valence (2002, p. 76)

The choice of building materials affects the environmental impact of the construction of a masonry building as all building materials are processed in some way before they can be used in the construction process. The greater the number of processes a material or set of components have to go through, the further a material has to travel and the heavier the material is, the higher will be its embodied energy and the number of associated waste products. Haulage of construction materials within the UK accounts for some 20% of the embodied impact of construction.

In a typical house built with brick and concrete block cavity walls, it has been suggested that the masonry will account for around 30% of the total energy embodied in its materials – the highest total for any construction element (Roaf et al, 2004). However, it is extremely difficult to find accurate published figures for the energy embodied in construction materials. The fact that estimates of energy embodied in materials as a proportion of total energy use can vary so vastly between 2% (Lazarus, 2002) and 24% (Best and de Valence, 2002) bear out this fact. Typical levels of CO_2 embodied in masonry materials due to the energy expended during their extraction/ manufacture are detailed in Table 2.

Table 2: **Typical levels of CO_2 embodied in masonry materials**

Material	Embodied carbon (kg CO_2/kg)
Cement	0.74 (manufactured in dry kiln)
Clay bricks	0.3–1.4
Concrete bricks/blocks	0.28–0.375
Calcium silicate bricks	0.13–0.25
Stone	0.012 (limestone)–0.32 (granite)
Mineral/glass wool insulation	1.2–1.35
Cellular plastic insulation	3.4

It is not the author's intention to offer detailed information on a subject that, being a particularly new science, could form the basis for a substantial publication in its own right. The following information merely considers how the manufacturers and users of each of the masonry materials can give consideration to the reduction of the energy associated with their production and use, while at the same time taking into account using recycled content in their makeup, facilitating their eventual reuse and standardising material dimensions to reduce wastage and overall durability.

3.3 MORTAR

Mortar is one of the most basic elements of masonry construction, although its functions are numerous:

Mortar is a mixture of materials for jointing masonry units. It sticks the bricks together to provide stability and solidity while holding them apart to spread loads evenly. It compensates for irregularity between units when straight, level and plumb walling is laid. It also seals any gaps to resist wind or rain penetration. As well as fulfilling its gap-filling adhesive function it is required to have durability and strength to suit the application. The physical properties of a mortar depend on the nature and proportions of its constituents.

Mortar Industry Association (1988, p. 22)

In terms of the sustainability of masonry mortars, their production and use should ideally involve:
- Use of the least amount of energy in binder production as is practically possible, thereby reducing resultant CO_2 emissions
- Use of alternative fuels during the burning process in the production of cement or other binders
- Facilitation of the eventual reclamation of masonry units by ensuring bond strength that is not excessive (see Chapter 4, 'Design for deconstruction and reuse')

- Premixing with aggregate and delivery to site in silos to ensure consistency and reduce site wastage

Realising the ever-increasing emphasis that the government is placing upon sustainability issues, advocates of hydraulic lime mortars often portray the material as a 'green' alternative to cement, claiming that the production of hydraulic lime uses less energy and, subsequently, releases less CO_2 into the atmosphere. It is also claimed that the pure lime (or free lime) contained within hydraulic lime mortars reabsorbs the volume of CO_2 driven off during its production as it sets or carbonates.

Table 3, taken from British Standard BS 5628-1 (BSI, 2005a), groups mortar mixes within four designations. Within each designation, mixes produce mortars of approximately equal strength and durability. As detailed in Chapter 2, BS 5628 does not cover the use of lime mortars. Table 4, showing similar information for natural hydraulic lime mortars, is taken from *The use of lime-based mortars in new build* (NHBC Foundation, 2008), which contains a Draft for Development Standard for lime mortars that is intended to be used in conjunction with BS 5628-1.

Hydraulic lime binders take up to three times as long to reach the compressive strengths required of cement mortars by British Standard BS 4721 (*Specification for ready-mixed building mortars*) (BSI, 1981). Therefore, the 28-day strengths achieved by cement-based mortars

Table 3: Requirements for mortar

Mortar designation	Type of mortar (proportion by volume)			Mean compressive strength at 28 days (N/mm²)	
	Cement: lime:sand	Masonry cement: sand	Cement: sand with plasticiser	Preliminary (laboratory) tests	Site tests
(i)	1:0–¼:3	-	-	16.0	11.0
(ii)	1:½:4–4½	1:2½–3½	1:3–4	6.5	4.5
(iii)	1:1:5–6	1:4–5	1:5–6	3.6	2.5
(iv)	1:2:8–9	1:5½–6½	1:7–8	1.5	1.0

Increasing strength ↑

Increasing ability to accommodate movement, eg due to settlement, temperature and moisture changes ↓

Direction of change in properties is shown by the arrows

Increasing resistance to frost attack during construction →

Improvement in bond and consequent resistance to rain penetration ←

are measured at 91 days for hydraulic lime mortars. This would seem to present a problem in relation to the progression of consecutive lifts of masonry over short periods of time, lifts of up to 1.5 m per day being possible under British Standard BS 5628-3 for cement mortars (*Code of practice for use of masonry – Materials and components, design and workmanship*) (BSI, 2001a).

Since their reintroduction as binders for low-rise new-build projects, hydraulic lime mortars have, in the main, been limited to buildings with a fairly large footprint (eg Haberdashers' Hall in London and the RSPB headquarters in Bedfordshire). In such instances, it may be possible to programme works so that they progress around the building from a particular starting point. Having been involved in the design of the RSPB headquarters, Beare (2004) claims that masonry built with hydraulic lime can progress as quickly as with cement-based mortars on this basis. However, during previous discussions with the author, Ian Pritchett of Lime Technology Ltd has claimed that, in his experience, lifts of over 1.5 m high are possible with most projects when using hydraulic lime mortars if suitably porous masonry units are specified. It seems clear that research into the possible speed of construction of conventional masonry buildings using hydraulic lime mortar is needed to instil confidence in its use.

The techniques used to produce hydraulic lime and cement and their resultant CO_2 emissions are now considered in detail.

Table 4: **Natural hydraulic lime mortars for use with masonry**

	Compressive strength class	Prescribed mortars (proportion of materials by volume)*			Compressive strength at 91 days (N/mm^2)†	Site-tested compressive strength at 91 days (N/mm^2)
		NHL 2 Natural hydraulic lime: sand	NHL 3.5 Natural hydraulic lime: sand	NHL 5 Natural hydraulic lime: sand		
While all natural hydraulic mortars will accommodate movement, increasing the mortar designation will decrease the ability to accommodate movement, eg due to settlement, temperature and moisture changes	HLM 5	-	1:1	1:2	5.0	4.0
	HLM 3.3	-	1:1½	1:2½	3.5	2.5
	HLM 2.5	-	1:2	1:3	2.5	1.5
	HLM 1	1:2	1:3	-	1.0	0.5

* Proportioning by mass will give more accurate batching than proportioning by volume, providing that the bulk densities of the materials are checked on site.
† The compressive strength at 28 days would be expected to be half of these values.

3.3.1 Production of hydraulic lime

Calcium carbonate (limestone) is formed in sedimentary rocks deposited over millions of years by the decay of sea life. Chalks, carboniferous limestone and rocks of the Jurassic period are the main sources. Calcium limes (also known as fat limes or air limes) are limes containing extremely low levels of impurities such as clay (levels as low as 5%). These limes result from limestone heated at temperatures of around 900°C. Carbon dioxide is driven off, leaving the highly reactive mineral calcium oxide or quicklime, which then reacts vigorously when mixed or slaked with water to produce calcium hydroxide. When calcium hydroxide is used as a binder for mortars, the free lime (pure lime without impurities) reabsorbs the CO_2 driven off during its production, and eventually the lime returns to a state that is similar to the parent limestone (Holmes and Wingate, 2002). In some cases, this 'induration' process (starting along the exposed surfaces of the mortar joints) can take hundreds of years to complete, depending upon the width of the masonry (Wingate, 1987).

Calcium limes are supplied to the construction industry in either putty or hydrate (powder) form. Figure 11 shows the production and carbonation processes, often referred to as the 'lime cycle', in schematic form.

Lime is classified according to its ability to set under water. A formal classification system was introduced by Louis Vicat, an eminent French civil engineer, and as a result of his work (Vicat, 1837, reprinted 2003), limes that set under water are known as hydraulic limes, their former name being 'water limes'.

The modern term 'hydraulic lime', as used in British Standard BS EN 459-1 (*Building lime – Definitions, specifications and conformity criteria*) (BSI, 2001b), refers to three groups of products. Natural hydraulic limes (designated NHL) are limes produced by

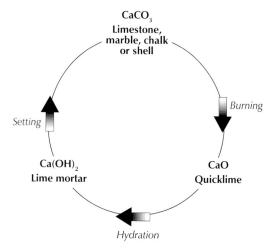

Figure 11: The lime cycle

burning argillaceous or siliceous limestones, with reduction to powder by slaking, with or without grinding. Special natural hydraulic limes (designated NHL-Z) are produced by blending natural hydraulic limes with up to 20% of suitable pozzolanic products (eg pulverised fuel ash or PFA, volcanic ash and trass) or hydraulic materials (eg ordinary Portland cement and blast furnace slag). Artificial hydraulic limes (designated HL) consist mainly of calcium hydroxide, calcium silicates and calcium aluminates and are produced by blending suitable powdered materials, such as natural hydraulic limes, fully hydrated air limes, dolomitic limes, PFA, volcanic ash, trass, ordinary Portland cement and blast furnace slag.

Many specifiers have, in recent years, stipulated the use of hydraulic lime mortars, believing the limes being delivered to site to be natural when they have in fact been one of the many available artificial products. These artificial products are usually imported from France and Italy and it may be said that they are not hydraulic limes at all, falling more suitably under the classification of masonry cements (Mortar Industry Association, 2004a). Natural hydraulic limes are only produced by one small works in the UK (Hydraulic Lias Limes Ltd of Dorset). Most of the natural hydraulic limes used in the UK are also imported from France and Italy (Mortar Industry Association, 2004a).

The processes required to produce natural hydraulic lime are very similar in principle to those for calcium limes, but due to the varying levels of impurities contained within the limestone being heated, they are slightly more complex. The burning process produces calcium oxide as well as calcium silicates and calcium aluminates, which are responsible for the hydraulic properties of the material. Variations in the raw material, as well as in the firing temperature, can produce hydraulic limes of very different characteristics.

Natural hydraulic limes have traditionally been classified by the amount of active clay materials they contain, as indicated in Table 5 (English Heritage, 1997). The degree of hydraulicity of the lime affects many of its characteristics when used as a mortar binder, eg compressive and tensile strength, speed of set in the presence of water, water vapour permeability.

After burning, the material used to produce natural hydraulic limes is subjected to a hydration process. Ideally, this process should just involve adding sufficient water during slaking to enable the conversion of the quicklime to powder without causing the calcium silicate components to start to set. In the case of eminently hydraulic

Table 5: Classification of hydraulic limes

Classification	Active clay materials
Feebly hydraulic (NHL 2)	< 12%
Moderately hydraulic (NHL 3.5)	12–18%
Eminently hydraulic (NHL 5)	18–25%

limes, the quicklime may need to be subjected to a grinding process prior to hydration. The resultant product is a hydraulic lime powder, which is delivered in sealed bags. As is the case with Portland cement, it is essential that the material be kept sealed and stored in a dry place, as any contact with water or atmospheric moisture can initiate the setting process (English Heritage, 1997).

Most of the articles and publications promoting the environmental benefits achieved in the production of hydraulic lime state that kiln temperatures required during its production are between 900°C and 1100°C. These temperatures are then often compared with those claimed to be required to produce cement (between 1400°C and 1500°C), the higher temperatures requiring the use of more energy/fossil fuels and therefore throwing more CO_2 into the atmosphere. It is often stated that the production of one tonne of cement releases one tonne of CO_2 into the atmosphere (Harrison, 2004).

English Heritage (1997) states that kiln temperatures required for the production of hydraulic limes (not just lime, as is often stated) can be as high as 1200°C. Ashurst (1997) and Oates (1998) state that temperatures as high as 1250°C are required, depending upon the clay content of the lime being burned. The fact that more eminently hydraulic limes require far higher temperatures during production than calcium limes is something that does not always seem to be taken into consideration when claims are made that the temperature differences between the production processes of hydraulic lime and cement are vast.

In relation to the claim that more energy is used by cement production (thus using more fossil fuels and creating more CO_2), BDA (2001, p. 4) states that:

Although kiln temperatures are lower for lime burning than for Portland cement production, lime kilns tend to be much less efficient. So this claim may only be theoretical.

It is also often claimed that lower CO_2 emissions can be achieved with hydraulic lime binders via reabsorption of CO_2 by the free lime within the mortar following use. However, there is currently no evidence that can substantiate claims that this is of substantial proportions. Carbonation is mentioned by the Foresight Lime Research Team (2003), but only to the effect that the purer the lime (ie calcium limes), the quicker the carbonation. The carbonation of calcium lime mortars, which contain few impurities, is well established. However, the more complex chemistry of hydraulic limes leaves any claims of substantial carbonation of these binders as purely theoretical, as is concluded by BDA (2001).

Figure 12 shows the chemical process in the production of natural hydraulic lime and highlights the percentages of the final constituents of the material. It can be seen that the amount of pure lime reduces as the production process progresses, leaving less free lime to carbonate.

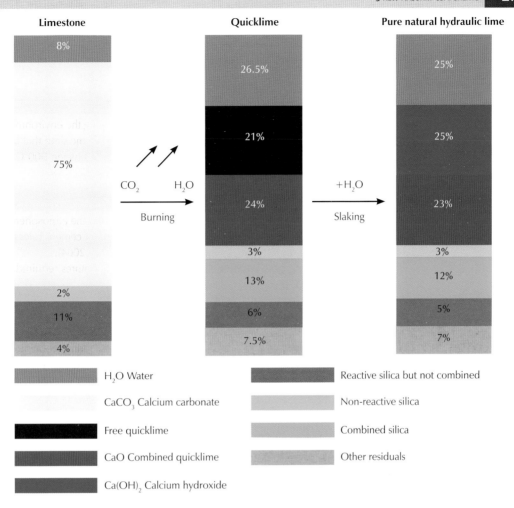

Figure 12: Chemical process in the production of natural hydraulic lime (based upon a binder classification NHL 3.5)

3.3.2 Production of cement

Portland cement is manufactured by combining a homogenous blend of carefully proportioned raw materials comprising lime (CaO), silica (SiO_2), a small proportion of alumina (Al_2O_3) and generally iron oxide (Fe_2O_3) at between 1400°C and 1450°C in a rotary kiln (CEMBUREAU, 1999). The raw materials fuse together to form 'clinker', a hard granular material. Clinker is ground to a powder along with gypsum to make cement.

Portland cement contains four main compounds (Hewlett, 2001):

- *Tricalcium silicate (C_3S)* – this reacts rapidly with water producing relatively large amounts of heat to form calcium silicate hydrate. It has a high strength and is the main contributor to the early strength of cement hydrate.
- *Dicalcium silicate (C_2S)* – this reacts slowly with water to form the same product as tricalcium silicate. It increasingly contributes to strength at later ages.

- *Tricalcium aluminate (C_3A)* – this reacts very rapidly with water.
- *Tetracalcium aluminoferrite (C_4AF)* – this reacts rapidly with water but does not produce much heat or strength.

Historically, the development of the clinker manufacturing process has been characterised by the change from wet to dry systems, with the intermediate steps of the semi-wet and semi-dry process routes. The first rotary kilns introduced by Frederick Ransome in 1885 were long wet kilns (Blount, 1920).

The main advantage of a modern dry process over a traditional wet system is the far lower fuel consumption. As the use of the dry process has increased, the energy required to produce cement has decreased by 30% in the UK and the rest of Europe since the 1970s (CEMBUREAU, 1998).

The four different basic cement production processes can be briefly characterised as follows (CEMBUREAU, 1999):

- *Wet process* – the slurry feed is dried and calcined within the kiln (conventional wet process); or fed to a slurry drier prior to a pre-heater/pre-calciner kiln (modern wet process).
- *Semi-wet process* – the slurry is dewatered in filter presses and the resulting filter cake is either:
  - extruded into pellets and fed to a travelling grate pre-heater; or
  - fed directly to a filter cake drier for (dry) raw meal production prior to a pre-heater/pre-calciner kiln.
- *Semi-dry process* – dried ground material (raw meal) is nodulised with water, then dried and partly calcined in a grate pre-heater; or in some cases, fed to a long kiln equipped with internal cross pre-heaters.
- *Dry process* – the raw meal is pre-heated in a series of cyclones (four or five stages), possibly incorporating a pre-calcining stage in which some of the fuel is burned or, in some cases, fed to a long dry kiln with internal chain pre-heater.

The specific energy requirements of the different kiln systems generally decrease significantly from the wet to the dry process, as do the specific amounts of most pollutants released. Table 6 shows the energy requirements of the different processes (Environment Agency, 2001).

Around 80% of Europe's cement production is from dry process kilns, and this percentage is growing constantly. Therefore, this process is considered now in isolation, as the wider discussion of all of the processes involved in manufacturing cement could in itself form the basis of a substantial body of work.

Dry kilns have been improved using two basic variants: pre-heater kilns and pre-calciner kilns. Cyclone pre-heater systems have been developed that carry out the pre-heating process outside the rotary kiln. The principle of all cyclone pre-heaters is that the raw meal is introduced and moves counter to the upward-flowing hot gases leaving the kiln. The meal is swept up in the gas stream, separated

Table 6: **Typical energy consumption of different kiln systems**

Kiln system	Specific fuel consumption (MJ/tonne clinker)
Wet (conventional)	6000–6500
Dry process long kiln	Up to 5000
Modern wet and semi-wet (pre-heater and pre-calciner)	4000–4800
Semi-wet (grate pre-heater)	3700
Semi-dry (grate pre-heater)	3300
Dry (pre-heater)	3500–4000
Dry (pre-heater and pre-calciner)	2900–3200
Theoretical heat of reaction	1700–1800

out in the cyclone and passed down to the next stage where the procedure is repeated until the raw meal has passed through all stages, rising rapidly in temperature as it does so. Heat transfer within cyclones is very efficient when compared with the kiln. The contact area with the finely divided raw meal is much greater than with a bed of material lying in the kiln, with a limited surface contact area exposed to the hot gases (CEMBUREAU, 1999).

The most modern and efficient variant of cement kiln technology involves pre-calcination, where decomposition of calcium carbonate in the kiln feed is achieved in suspension, using a separately fired pre-calciner furnace (Figure 13) feeding into a relatively short rotary kiln. In such an arrangement, approximately 60% of the fuel required is burned in the pre-calciner, thereby ensuring that 80–90% of the calcination of the raw meal takes place. Thus, when the meal enters the kiln, only final calcinations and clinkering are required (CEMBUREAU, 1999).

The cost of this technology is not cheap. Cemex invested £150 million in a new pre-calciner kiln arrangement at its Rugby works. Cement producers such as Cemex manufacture and sell millions of tonnes of their products per annum and costs associated with technological development can be absorbed within the substantial finances being generated by the company. In comparison, St Astier Limes of France, one of the main exporters of natural hydraulic lime to the UK, produces only 100 000 tonnes of its products per annum (Setra Marketing Ltd, 2001). It is clear that with such small production rates, major investment in kiln technology is out of the reach of natural hydraulic lime producers.

Figure 14 shows a typical pre-calciner dry process flow diagram (Environment Agency, 2001).

Figure 13: A pre-calciner tower, raw meal silo and exhaust stack

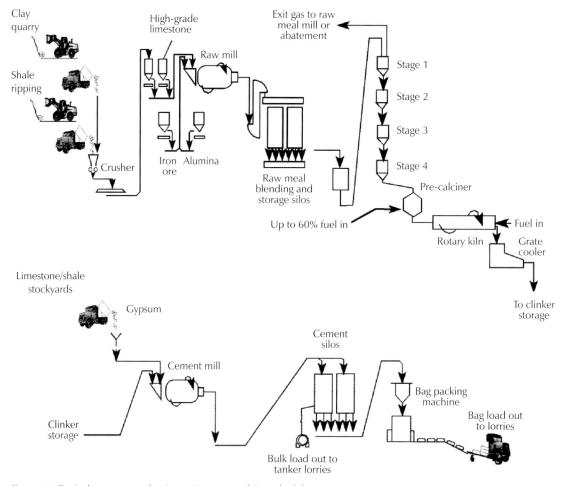

Figure 14: Typical cement production using a pre-calciner dry kiln

Far from releasing one tonne of CO_2 into the atmosphere for every tonne of cement produced – a statistic that is often put forward by advocates of lime mortars – the use of pre-calciner dry kiln technology has allowed cement manufacturers to cut energy levels by one-third, thereby reducing CO_2 emissions to around 700 kg per tonne of cement.

UK cement producers have also implemented the use of alternative fuels containing waste products that would otherwise be sent to landfill sites in an attempt to reduce the use of fossil fuels and, therefore, subsequent emissions of CO_2. Castle Cement has developed its own alternative fuel, named Profuel. The waste used is non-hazardous and includes such materials as paper, cardboard and packaging, normally too expensive or difficult to recycle. Through the development of Profuel, paper, plastics and other fibrous wastes can be shredded and mixed to a recipe that can be fed into a cement kiln. Because of the extremely high gas temperatures of a kiln, the material is totally burned. Utilisation of Profuel is authorised by the Environment Agency and its use saved 15 400 tonnes of CO_2 emissions in 2002. It has been estimated that the 15 existing cement plants in the UK have the capacity to treat approximately 1.5 million tonnes of waste in this way annually (House of Commons (online)). The average cost of cement production is around half that of hydraulic lime.

3.3.3 Magnesium oxide cements

Currently under development, magnesium oxide cements are derived from two distinct chemical/mineralogical forms: magnesium carbonates (eg the mineral magnesite) or magnesium silicates. In each case, the cementing substance will be reactive magnesium oxide. As is the case with the production of Portland cement, large quantities of CO_2 are driven off when magnesium carbonates are heated, but with magnesium silicates there is no chemically bound CO_2 to be emitted. 'Eco-cements' derived from magnesium carbonates have been invented in Tasmania, but those derived from magnesium silicates have created more interest in the UK due to the country's lack of magnesite deposits (BCA, 2009).

Novacem, a spin-off company of Imperial College London, has developed a magnesium oxide cement derived from silicates, which, the company claims, can absorb more CO_2 (1.1 tonnes) than is emitted into the earth's atmosphere during its production (0.5 tonnes).

However, until such products become commercially available and reliable data can be gathered over time in use, their viability in terms of durability and the facilitation of the reuse of masonry units will not be known.

3.3.4 Aggregates

The performance of a mortar is heavily dependent upon the quality of the aggregate which should, under ideal conditions, be sharp in particle profile and well graded through the selected range of sizes.

Bennet (1999, p. 3)

The choice of aggregate is especially essential to the formation of a successful hydraulic lime mortar mix. The best mortar sands generally have a void ratio of around 33–35% (Historic Scotland, 1998), which means that they will require the addition of around one-third of their dry compacted volume in binder paste in order to fill all the voids. This is the basis of the traditional 1:3 binder:aggregate recipes for modern mortars. Livesey (2002, p. 29) sums up the requirements of sands for hydraulic lime mortars:

Hydraulic lime mortars require aggregates with different properties to those used for cement mortars. Clean, sharp sand is preferred, having a size range such that the grading is evenly spread over a number of size fractions. The softer cement mortar sands would give lime mortars with high water demand, and tend [to] be over-cohesive due to their fines content.

Under British Standard BS 1200 (*Specifications for building sands from natural sources*) (BSI, 1976), there are two sand types and grading requirements for cement mortars: type S and type G. The Mortar Industry Association (2004b, p. 5) expresses concerns about these requirements:

It should be noted that whichever compliance system is used it is still possible for sand to comply with the standard but to be deficient in some respects. Certainly in respect of the perceived workability of the mortar, it is possible for complying sands to nevertheless be excessively single sized with [a] concomitant tendency to bleed and segregation.

Here, two observations can be made:
- The aggregate demands of hydraulic lime mortars are stricter than those for cement mortars.
- More attention needs to be paid to aggregates specified for cement mortars. The specification of sand is less essential for these mortars, but ensuring that well-graded sand is used could ensure vast improvements in their quality.

In terms of the most sustainable options for aggregate, although research has been carried out into the use of aggregate containing recycled glass with lime binders (Gervis, 2004), product development has not yet proved fruitful and the only option would appear to be the continued extraction of building sands. For the purposes of functionality, well-graded building sands should be used for masonry mortars, especially with lime-based binders, although this is not always found to be the case. Research by Yool and Lees (1995) found that 40 of a sample of 154 building sands taken from all over the UK did not comply with any British Standard for building sands.

3.3.5 Factory-produced mixes

Factory-produced mixes now account for over 75% of all mortar used in the UK each year, a sure sign that the construction industry in this country is attempting to move away from the poor batching practices that have been encountered on sites for many years.

Each production plant takes in various sands and dries them in their kilns before elevating them into the roof-top storage silos. Each mortar recipe is stored in a computer and batched by weight into the mixer, where it is thoroughly mixed before being discharged into a silo (or tanker). The silo (Figure 15) is then delivered to site on the back of a purpose-built lorry. It is hooked up to the electricity and water supply and fully commissioned by an engineer. Water content of the mortar produced by the mixer at the base of the silo is computer controlled, ensuring product consistency. The silo will contain up to 15 tonnes when it is delivered, and is then refilled by a tanker, which pumps the mortar directly into the silo. The tanker can deliver up to 30 tonnes at a time. The silo can be refilled on a regular basis until the job is complete and the silo is collected.

Figure 15: Mortar silos

A small hopper-style system can also be offered for smaller sites, with pre-mixed bags being emptied into low hoppers, which replace the silo above the base mixer. The requirements of British Standard BS EN 998-2 (*Specification for mortar for masonry – Masonry mortar*) (BSI, 2003a) for factory-produced mortars will hopefully continue to hasten the already rapid demise of mortars produced on site, as the Standard allows mortars to be either prescribed or designed and they can be CE marked. The fundamental change that BS EN 998-2 has introduced since 2005 is a requirement that specifications for factory mortars are performance led, the Standard being based on the compressive strength of the set mortar, thereby ensuring consistency.

3.4 MASONRY UNITS

Time has already demonstrated beyond doubt that masonry (particularly clay brickwork) is the most durable of all of the building materials available. Alongside timber, it is the oldest of the building materials and there are many fine examples throughout the world of its excellent durability qualities, often enduring severe exposure conditions over many centuries.

Parkinson et al (1996, p. 63)

3.4.1 Functionality

Structural masonry units should comply with the following relevant British Standards (where applicable, the replacement BS EN Standards are shown in parentheses):

- Calcium silicate (sandlime and flintlime) bricks: BS 18799 (BS EN 771-2; BSI, 2003b)
- Clay bricks: BS 3921 (BS EN 771-1; BSI, 2003c)
- Clay and calcium silicate modular bricks: BS 6649 (BSI, 1985)
- Dimensions of bricks of special shapes and sizes: BS 4729 (BSI, 2005b)
- Natural stone masonry: BS 5628-3 (BS EN 771-6; BSI, 2005c)
- Reconstructed stone masonry units: BS 6457 (BS EN 771-5; BSI, 2003d)
- Precast concrete masonry units: BS 6073-1 (BS EN 771-3 and BS EN 771-4; BSI, 2003e and 2003f)

The way a masonry unit is made affects its robustness, colour, shape, strength and texture. Therefore, it is essential that designers of masonry structures understand enough about manufacturing processes to know what to ask for and to appreciate the limitations of the processes involved. Before a masonry unit is chosen, it might be prudent to obtain evidence of performance data from previous projects where those units have been used. Test results of recent production runs might also be useful or, where no recent test certificates are available, tests to demonstrate that the units satisfy specific engineering requirements may be applicable (Institution of Structural Engineers, 2005).

The manufacture and use of each of the main types of masonry unit are now considered separately.

3.4.2 Clay bricks

Building with brick is building in the most direct sense and is in harmony with the environment. Brick reflects the local geology, so it is so satisfying to tour an area and see the colour of the soil reflected in people's homes; it gives direct ties with a location.

Shuttleworth (2006, p. 5)

The dominant material for brick making in the UK is clay, with nearly 90% of all the bricks used being clay bricks (Hammett, 1991). The acceptability of the built environment is largely dependent upon its aesthetic appeal. Clay facing brick makes a significant contribution in this respect because of its wide range of colour and texture derived from the wide variety of clays and production techniques. The plethora of available products allows its use in a wide variety of styles, both by itself and in conjunction with other materials (Figure 16).

Figure 16: Samples of the many available clay brick colours and textures

Clay is formed by the grinding and disintegration of rock and its geology is complicated, the term 'clay' covering a large number of different substances that share similar properties. By mixing with water, clay becomes workable, this being a change in the material that is reversible. However, when the clay is fired, all water is removed and this change becomes irreversible. Water cannot then be reintroduced into the clay as it has become a ceramic material. There are three techniques used to form bricks (BDA, 1999):

- *Soft mud process* – a free-flowing clay mix with up to 30% moisture content is thrown into mould-boxes either by hand or machine then dried and fired.
- *Extrusion process/wire cut* – clay is forced by an auger through a lubricated die to form a continuous column of stiff clay, which can be 'faced' by roll-texturing, sand-blasting and pigment-spraying to produce a range of textures and other aesthetic effects. The column is 'wire cut' into bricks using tightly strung steel wires.
- *Pressing* – semi-dry clay is pressed into a mould-box.

Clay is an abundant resource and the production process is relatively free of harmful by-products. However, the firing process gives clay bricks relatively high embodied energy levels. During Roman times, clay bricks were fired at temperatures no higher than 350–450°C (Berge, 2000) and many of the structures built with these materials still stand today (Figure 17). In order for a modern clay brick to vitrify, it must be heated in a kiln to a temperature between about 900°C and 1150°C (Figures 18–21), this heat being maintained for at least 8–15 hours with the exact temperature dependent on the type of clay used (Campbell and Pryce, 2003).

Figure 17: The Aqua Claudia, one of the first aqueducts in Rome, completed in 47 AD

Figure 18: Waste waiting to be crushed to 'grog'. This picture and the following photos (Figures 19–21) were taken at Furness Brick, which is an independent, family-run brick manufacturer operating near Barrow in Cumbria. As well as the dark virgin clay that is acquired within a radius of a few miles of the Furness Brick plant, waste products waiting to be crushed down to 'grog' and included as constituents of new clay bricks can be seen here in the foreground.

Figure 19: Having been pressed from crushed material to which water has been added, these clay bricks are on their way to be fired

Figure 20: Here, freshly pressed bricks are being stacked in a traditional brick kiln, a process that has changed little since Furness Brick was formed in 1845

Figure 21: Recently fired clay bricks. The dark grey bricks in the foreground are still to be fired. The results of traditional kilns are a unique range of products that display a variation of appearance, quality and charm rarely found in modern bricks.

Some clay brick manufacturers are now seeking to improve their profile in terms of environmental awareness, while at the same time attempting to lower rapidly increasing production expenses brought about by recent steady rises in fuel costs. Such companies are now beginning to realise that long-term financial and environmental benefits can be gained by investing in new production techniques and reducing the drain on fresh non-renewable resources. Table 7 details the main ways in which clay brick manufacturers have sought to reduce the impact that their activities have on the environment.

Table 7: **Environmentally friendly moves made by clay brick manufacturers**

Activity	Improvements
Extraction of clay	• Clay extraction has a disruptive and adverse environmental impact. However, subsequent restoration of clay pits often adds value through the provision of leisure facilities and areas dedicated to wildlife and nature conservation. Restoration of clay pits can also provide land for agricultural and other productive uses. • Additives are included in the clay mix, thereby reducing the amount of virgin clay extracted. Waste materials such as manufactured waste, colliery spoil, dredged silt, PFA, steel slag, blast furnace slag and sewage sludge can all be used to replace clay content (CIRIA, 1999a). Increasing the recycled content of building materials also diverts waste from landfill, generating a market for materials that would otherwise cost money. Larger companies have set a target to produce products with 5–10% waste content at each of their factories. Products containing waste materials have been proved to possess equal properties to those made totally from clay. • Energy consumption is also related to transport needs – haulage of construction materials within the UK accounts for some 20% of the embodied impact of construction. Fired clay products are heavy, and industries producing them are relatively centralised. Some manufacturers are attempting to ensure that raw materials are sourced locally, the clay being taken out of the ground in quarries adjacent to manufacturing plants.
Firing of clay	• De Vekey (1999) suggests that the overall energy input to all brick-manufacturing processes fell by around 20% between 1974 and 1996. Ibstock Brick Ltd set a target of a further 10% reduction in energy use by 2010, having trimmed 20 000 tonnes from its annual CO_2 emission rates between 2004 and 2005 due to improved kiln efficiencies (Ibstock Brick Ltd, 2006a). • The move to more highly perforated bricks minimises both resource usage and energy use, due to the reduction of material to be fired. However, attempts to reuse these units in the future would be made difficult due to the fact that it would be hard to remove the mortar that gathers in the voids inside the bricks. • The use of alternative fuels, whether landfill methane or derived from toxic waste, is currently a popular environmental option with manufacturers. Manufacturers are exploring the use of landfill gases to generate electricity on those sites where the clay quarry is available for commercial landfill activities that are operated by third parties. • Companies have introduced unfired bricks to the market for use in non-load-bearing internal walls. One such product is Ibstock Brick Ltd's Ecoterre™ earth brick, which is fired with low-grade recycled kiln heat, thus reducing energy requirements for fully fired products that would otherwise be used for internal partition walls (Ibstock Brick Ltd, 2006b).

Although the recent introduction of unfired products such as Ibstock Brick Ltd's Ecoterre™ earth brick (also available as a block, see Figure 22) is admirable in terms of attempting to reduce the impact that the production of masonry units has on the environment, reclaimed masonry units could reasonably serve the same purpose for use in non-load-bearing masonry. It would seem that any concerns that specifiers may normally have in terms of the unknown characteristics of reclaimed materials would not be relevant in such instances.

100% of masonry and precast concrete can be recycled or reused at the end of the structure's life.

MMA (2006, p. 4)

To this point in time, the primary demand for reclaimed masonry is for clay bricks (Addis, 2006; CIRIA, 1999a). Despite the claim made above by the Modern Masonry Alliance (MMA), it is perhaps surprising that no current trade literature published by clay brick manufacturers makes mention of any type of consideration towards the eventual reclamation of their products. Research by Ribar and Dubovoy (1988) suggests that the surface texture of a brick is a primary factor in the development of bond strength in masonry. Might it not be prudent to examine this theory and consider ways in which to make new masonry units that would be easier to clean up on reclamation?

Figure 22: Ibstock Brick Ltd's unfired Ecoterre™ earth brick/block

3.4.3 Calcium silicate bricks

Calcium silicate bricks (sometimes referred to as sandlime or flintlime bricks) are made by mixing together hydrated lime, sand (possibly with the addition of crushed flint) and a pigment for colour (if required) with enough water to enable the mix to be moulded. The units are then cured with saturated steam in an autoclave (essentially a large pressure cooker) at around 200°C (Figures 23–25). The process was invented and patented in England by Van Derburg in 1866. This was then developed and patented on 5 October 1880 in Germany by the German researcher Dr Wilhelm Michaëlis, but calcium silicate was never anywhere near as popular as clay in the UK. The widespread development of concrete masonry units in the latter part of the twentieth century led to the even more limited use of calcium silicate products.

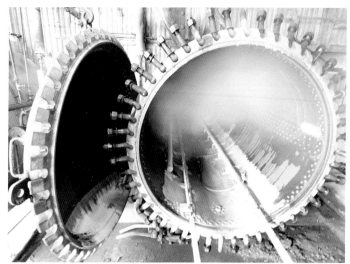

Figure 23: Autoclave used to manufacture calcium silicate bricks

Figure 24: Side view of autoclave

Figure 25: Uncured calcium silicate bricks waiting to be moved into the autoclave

Figure 26: Samples of calcium silicate brick colours and textures

However, as little as 40% of the energy required to manufacture clay products goes into the creation of calcium silicate bricks and, as is the case with clay bricks, it is possible to include waste materials such as blast furnace slag, china clay sand and spent oil shale as a percentage of the matrix used to manufacture calcium silicate bricks (CIRIA, 1999a). Rejected products can also be recrushed and used in the production of new bricks. In terms of impact on the environment, calcium silicate bricks are an attractive option.

A natural calcium silicate brick is near white when white sand is used in its production and a little pink if red sand is used, but with the use of pigments and differing casting techniques, like clay products, a wide variety of attractive finishes are achievable (Figure 26).

As calcium silicate bricks have been manufactured and used for only a little over 100 years, reliable data with regard to their long-term durability (compared with clay and stone masonry units, which have a proved durability over many centuries) and eventual reclaimability are not available. Nevertheless, as it is widely recognised that the bricks harden with age due to their hydrated lime content, durability is unlikely to be an issue to raise concerns.

3.4.4 Concrete bricks/blocks

Concrete masonry units use Portland cement to bind an aggregate together, with the colour and density of the aggregate determining the colour, weight and performance of the unit (Figures 27–29). As they can be made lighter than fired clay products by using low-density aggregates, concrete units can be made larger, allowing walls to be built more quickly. It is primarily for this reason that concrete blocks first began to replace clay brickwork in the middle of the twentieth century.

Figure 27: Dense concrete blocks

Figure 28: Lightweight concrete blocks

Figure 29: Aircrete blocks

Lightweight aircrete blocks are now required more generally for their thermal insulation properties and regularly form the internal leaf of new cavity construction. However, some concrete block manufacturers are now offering masonry systems that can be used to complete whole buildings, H+H Celcon Ltd's 'Rå House' being one such system. The Rå House can be formed in either cavity or solid wall construction. The benefits of such systems, which are regularly quoted in trade literature (Celcon, 2005; Aircrete Bureau, 2005; Thermalite, 2005), are rapid build times, increased productivity and reduced costs, all very attractive to developers looking to squeeze the maximum returns from their investments.

Figure 30: Silos containing the constituents of aircrete blocks

Figure 31: Silo contents are fed into a mixer, where the slurry is prepared

Concrete blocks and bricks can compete with clay products in terms of energy savings as only the small proportion of Portland cement contained within their makeup requires high temperature treatment rather than the whole masonry unit (CIRIA, 1999a). The aggregates, which are relatively inexpensive, provide the bulk of the units and can contain waste materials such as PFA and blast furnace slag. In fact, PFA accounts for up to 80% of the raw material used in the manufacture of aircrete products (Exall, 2007).

Aircrete blocks are produced by mixing PFA and/or sand with cement and/or lime slurry, to which a small quantity of aluminium powder is added (Figures 30 and 31).

The slurry is poured into large steel moulds where a reaction between the aluminium powder and alkaline environment in the mix takes place, generating tiny air bubbles that stabilise to form the aircrete cellular structure (Figure 32).

Figure 32: Slurry setting in steel mould

Figure 33: 'Cake' being cut into blocks by tensioned wires

Figure 36: Cured aircrete blocks emerge from the rear of the autoclave

Figure 34: Sliced cake ready to be wheeled into an autoclave

Figure 37: Machines crack apart and shrink wrap the blocks

Figure 35: Front of autoclaves used to cure aircrete blocks

Figure 38: Packs of Thomas Armstrong airtec blocks awaiting transportation from the Catterick Plant

Over a period of around one and a half hours, the slurry rises and sets to form a 'cake', which is then cut by tensioned wires to produce the blocks (Figures 33 and 34). Any material that does not make it into the product is recycled into the next mix.

As is the case with calcium silicate bricks, aircrete blocks are cured in autoclaves under steam and pressure before the blocks are removed, packaged and ready for use (Figures 35–38). Steam generated during the curing process is either transferred to be used for curing the next autoclave of blocks or condensed and used in the process for making more mixes.

Due to the air cells generated in the mixture prior to steam curing, aircrete blocks comprise up to 60% air by volume, which gives the blocks their low weight. This in turn means that, in comparison with more conventional masonry materials, more blocks can be transported at any one time, thereby reducing exhaust emissions.

While waste materials such as silt and sludge (which are not derived from energy use) can be used in the manufacture of clay units, PFA and blast furnace slag (which can also be used in the manufacture of clay units) are derived from energy use. However, having been produced in vast quantities since the 1950s, PFA is readily available, with estimates that around 115 million tonnes of existing stockpiles can be accessed (Sear, 2007).

As is the case with calcium silicate bricks, reliable data with regard to the long-term (in terms of centuries) durability and eventual reclaimability of concrete masonry units are not available, although ancient crude concrete structures (such as the Pantheon in Rome) suggest that the durability of modern concrete masonry units should be considerable.

3.4.5 Stone

The UK has abundant examples of ancient structures that bear remarkable testimony to the longevity of stone as a building material, and because it occurs naturally, stone is enduring and attractive. Natural stone masonry was used extensively in the construction of buildings before concrete came into common use, but as masonry construction underwent considerable change throughout the twentieth century, so the use of stone for masonry construction diminished. Apart from the use of thin slabs as cladding panels and the occasional use of stone on prestigious building projects or where local vernacular tradition dictates (eg slate in the Lake District, Cotswold stone in parts of the Midlands, York stone in Yorkshire), artificial or reconstructed stone formed with concrete is often accepted as a stone substitute. However, while natural stones show an attractive variation in their colour and texture, similar characteristics can rarely be replicated in concrete, which tends to be more uniform in colour and texture. A major drawback against the selection of stone over facing brick is cost. For example, the cost of an external cavity leaf built with rough-dressed Cotswold stone is double that of facing brick walling

(Davis Langdon, 2007), and this cost difference becomes even greater when stones of a higher quality are considered. The use of stone is still prevalent in many countries with low and medium industrialisation, however, where it can cost as little as a quarter of the price of concrete (Berge, 2000).

There are three main categories of stone:
- Igneous or primary, generally granite, basalt, diorite or serpentine
- Sedimentary or secondary, generally limestone or sandstone
- Metamorphic or tertiary, generally marble or slate

Although the use of granite and slate (Figure 61, p. 74) is not unusual by any means, limestones and sandstones are the stones used chiefly for general building purposes, and there are long-term sources of these materials in the UK. Provided quarries are local and sufficient project funding will allow its use, stone is an obvious choice of material to use in terms of the overall impact its use has on the environment. Limestones (eg Cotswold stone, Portland stone, Bath stone) consist of particles of carbonate of lime dissolved together with similar material – generally, but not always, in sea water (back in geological time). Sandstones (eg York stone, Red St Bees) are composed of consolidated sand and consist chiefly of grains of quartz (silica) united by a cementing material. The quartz grains, dominantly between 0.07 mm and 2 mm in diameter, are practically indestructible, and are bonded together by calcite or quartz cementing material deposited between the surfaces of the grains.

Most natural stone is processed as follows (Ashurst and Dimes, 1990; DCLG, 2007a; BGS, 2005):

1. *Quarrying or mining the stone* – by drilling and plugging, manual cutting, percussion, hydraulic splitting, blasting and power sawing (Figures 39–43).
2. *Seasoning* – the age-old process of allowing the blocks of stone to stand for some time before use. Many stones can be worked by hand more easily immediately after quarrying, but because of the high water or 'quarry sap' content of some stone, which at this point is classed as being 'green', it needs to be seasoned first. Salts in the stone will tend to move towards the surface during seasoning and are removed when the stone is dressed.
3. *Preparation* – can be by hand working or by mechanical means. Soft stones can be shaped by hand or cut using hand saws; harder stones can be sawn using frame saws, diamond rotary blades/ wire saws or high-pressure water jets. Most building stones are prepared by mechanical primary sawing (Figure 44).
4. *Finishing* – can involve polishing using abrasives, texturing using a high-temperature flame jet that causes spalling, planing or, alternatively, milled/grooved surfaces can be created using computer-controlled routing machines, plain chisels or pneumatic hand tools (Figure 45).

Figure 39: Drilling of Red St Bees sandstone at Grange Quarry in Cumbria

Figure 40: Explosive charges being prepared for insertion into holes in the sandstone

Figure 41: Split sandstone following detonation of explosives

Figure 42: After plugging and further manual splitting, the stone is dragged up from the base of the quarry

Figure 43: Numbered stone awaiting transportation from the quarry

Figure 44: Profiling saw at York Minster stonemasons' yard. The saw is computer controlled and does much of the required shaping prior to finishing in the workshop.

Figure 45: Pneumatic hand tooling of York stone shaped by the profiling saw in the York Minster stonemasons' workshops

Other surface finishes include those of mechanical or hand tooling to obtain different surface effects (Figure 45). Finally, there is banker mason work for aesthetic or functional effects including profiling to shed rainwater from the building face.

The amount and type of work involved in preparing stone masonry units for the construction process is dependent upon the type of finish required. Historically, stone used as part of the often rendered rubble work used to construct some buildings prior to the onset of cavity construction required little or no preparation. In contrast, stone used for high-quality coursed ashlar work requires high levels of preparation due to fine jointing tolerances and, in some instances, special surface finishes (eg furrowed, punched, boasted).

Methods of stone quarrying and stone working have changed little in the course of time, with picks, crowbars and wedges still being used widely for extraction of sedimentary stones. Extraction is a mechanical process with no need for high temperatures and so there is very little embodied energy associated with the extraction of stone.

Quarrying is itself visually and ecologically damaging, although it is possible to locate new quarries where topography and vegetation provide screening in order to reduce negative impacts on amenity. Quarries can also create opportunities for geological and biological conservation (ODPM, 2004a). Under the Town and Country Planning (Environmental Impact Assessment) (England and Wales) Regulations 1999, Environmental Assessments are only required for proposed quarrying operations that fall within sensitive areas or exceed 25 ha in surface area. While there are just over 400 active stone quarries in the UK, the surface area of most building stone operations is well under 25 ha (DCLG, 2007a; BGS, 2005). Underground extraction is an alternative to quarrying where visible disturbance has to be avoided at all costs, but recovery rates/block sizes are restricted in comparison.

Stone is only a relatively low-energy product if it does not have to travel far to the point of use, as was the case historically. Prior to industrialisation and the mass production of masonry units, almost everything was built from the same local materials, which is why we have regional vernaculars (Smith et al, 1998). UK building stones often have very distinctive characteristics that cannot be matched by materials sourced elsewhere. However, the UK is a major and increasing net importer of low-cost stones, which are seen as the biggest threat to the indigenous building stone industry (DCLG, 2007a). The top five stone-producing and exporting countries are China, Italy, India, Iran and Spain, which together account for 74% of world production. Some stones from China and India can be sourced at costs that are 25% or less than alternatives extracted in the UK (BGS, 2005). Between 1996 and 2001, stone imports increased by 323% (ODPM, 2004b).

Since minerals, of any kind, can only be worked where they are found; and since building and roofing stones, in particular, depend so much for their appearance and importance on geological characteristics that vary considerably over short distances; the options for quarrying these materials in places that are convenient to other land use requirements and designations are often very limited. Planning for the supply of these materials has to take this as a fundamental starting point. If it does not do so, then the future production of indigenous stone, whether for the upkeep of historic buildings; maintaining vernacular styles of architecture in our towns and villages; or maintaining a viable building stone industry that can compete with imports from overseas, will be increasingly jeopardised.

ODPM (2004a, p. 6)

It seems clear that the UK planning process has a great part to play, both in maintaining vernacular styles through the use of local materials and in ensuring that unnecessary burdens are not placed upon the opening of new quarries.

Although imports have primarily had an effect on roofs, paving, kitchen worktops and internal refurbishment, there are signs that they are increasingly penetrating other sectors of the market, including that of stone for new build. Very few data exist in terms of the difference in levels of energy embodied in imported stone and that extracted in the UK, although common sense would suggest that this would, in all cases, be considerably higher in imports due to transportation alone (particularly for material sourced outside Europe). For example, Hammond and Jones (2006) claim that granite sourced from Australia contains more than double the carbon of that embodied in similar materials sourced locally (0.75 kg CO_2/kg versus 0.32 kg CO_2/kg).

Stone is also readily recyclable, as has been proved throughout time (Figure 46). Because it occurs naturally, stone is enduring and attractive.

Figure 46: Lanercost Priory in Cumbria, founded around 1169, was built with stone taken from nearby Hadrian's Wall, which had been constructed approximately 1000 years earlier

3.5 INSULATION

As air provides good resistance to heat flow, many insulation products are based upon materials that have numerous layers or pockets of air trapped within them. Some insulants reduce in efficiency if affected by moisture, so it is advisable to understand the properties of readily available insulants when specifying their use. The thermal efficiency of an insulant is denoted by its thermal conductivity (or lambda value), which is measured in W/mK and is the amount of heat transfer per unit of thickness for a given temperature difference.

The achievement of currently acceptable U-values for masonry wall construction is difficult without additional insulation. For example, a substantial solid wall made with a 100 mm facing brick skin, backed by 440 mm thick aircrete blockwork with a thermal conductivity of 0.18 W/mK and finished internally with 13 mm plasterboards on dabs, would achieve a U-value of 0.4 W/m²K, which falls short of the current industry target of 0.3 W/m²K. Only by increasing the blockwork width to 660 mm could the required U-value be achieved. Alternatively, merely adding an internal lining of a composite plasterboard with only a 20 mm backing of cellular plastic insulation would bring the solid wall up to the required standard. The standard could also be achieved by reducing the blockwork thickness to 215 mm and increasing the insulation thickness to only 50 mm. However, the German manufacturer Ziegel claims that its 365 mm wide cellular clay units, finished internally with a lightweight plaster and externally with a lightweight render, can achieve a U-value of 0.26 W/m²K without the use of insulation.

There are a wide range of insulation products on the market and the properties of each should be evaluated against the requirements of the job in hand. For example, does it need to be waterproof? How much room in the construction is available for it? How fire resistant does it have to be? Insulation is usually placed within the cavity between two leaves of masonry, or on the inside or outside face of a solid wall. By far the two most popular types of insulation now being used for masonry wall construction are fibre-based and organic/synthetic-based insulants (Smith, 2006). Both are considered in Table 8.

The placement of rigid insulants such as mineral fibre and cellular plastic in cavities must affect the potential durability of the masonry surrounding them due to the fact that the expected lifespan of the masonry will greatly exceed that of the insulation (XCO2, 2002). Injected loose full-fill cavity insulation systems have the potential to overcome this problem as they can be removed and replaced at a later date. However, a recent study by Kingspan (2000) has shown that such systems are not as energy efficient as the widely used options detailed above due to the fact that the formation of voids between the injected insulants cannot be properly monitored.

Part L of the Building Regulations (ODPM, 2006) is beginning to make both the partial and full filling of cavities increasingly difficult due to the government's move towards its ultimate aim of zero-carbon buildings. The 2006 version of *Approved Document L* lowered the required U-value for new external walls from 0.35 W/m²K to 0.30 W/m²K. The bar will continue to be raised (and the U-value therefore lowered), which will result in increased cavity widths that will be required to house more and more insulation, while at the same time, in the case of partial fill, retaining the required residual cavity of 50 mm. In terms of new housing, DCLG (2007b) has proposed requirements for a 25% improvement on 2006 levels of energy/carbon performance by 2010, followed by a further improvement of 44% on 2006 levels by 2013 before setting a zero-carbon requirement for 2016. In lieu of the great (and some may say unachievable) emphasis the government is placing upon the use of renewable energy sources to achieve future energy efficiency requirements, it is highly likely that a compromise will be struck by embracing the PassivHaus energy standard used in Germany (Hodgson, 2008).

Table 8: **Properties of insulation materials**

Insulant type	Properties
Mineral fibre 	• *Production* – Fibre-based insulants (such as Rockwool) are produced by melting a base substance at around 1600°C and spinning it into fibres, with a binder added to provide rigidity. They are recyclable at the end of their lifespan (XCO2, 2002). • *Thermal performance* – 0.033 W mK^{-1}. • *Use* – Usually used to fully fill cavities (see photo) or can be placed on the inside of solid walls. Mineral fibre cavity slabs contain fissures that direct moisture back towards the external leaf, thereby allowing full fill of a cavity. Embodied energy levels are 20% of those of some cellular plastic insulants (Lazarus, 2002). • *Lifespan* – Roaf et al (2004) estimate that the lifespan of mineral fibre is likely to be around 60 years. Knauf Insulation offers a 50-year guarantee on DriTherm cavity slabs to resist the transmission of liquid water from the outer masonry leaf to the inner masonry leaf in new external cavity walls (Knauf Insulation (online)).
Cellular plastic 	• *Production* – Produced by polymerisation using a blowing agent catalyst and surfactant. Not recyclable. • *Thermal performance* – 0.022 W mK^{-1}. • *Use* – Usually fastened to the internal leaf of cavity walls using proprietary wall tie clips, leaving a residual cavity of at least 50 mm, as required by Part C of the Building Regulations (ODPM, 2004c). Can also be used to insulate solid walls, both internally and externally. Roaf et al (2004) and Lazarus (2002) suggest that in terms of selecting insulation on the basis of minimum environmental impact, synthetic products should be avoided due to the high levels of energy used in their production, coupled with the fact that they are not recyclable. • *Lifespan* – Unknown, but unlikely to be anywhere near the potential lifespan of masonry construction. SIG Insulations (2009) claims that the thermal performance of phenolic foam insulant may be limited by the fact that its value is time averaged over 25 years. Kingspan offers a 25-year thermal performance guarantee on its panel insulation products (Kingspan Insulated Panels UK, 2006) but states that the durability of its cavity insulation boards has an 'indefinite life', which is dependent upon 'the supporting structure and conditions of its use' (Kingspan Insulation Ltd, 2007).

The principles behind PassivHaus are to construct a highly insulated and airtight envelope that will cut energy use to a minimum. Instead of heavily relying upon the use of renewable energy in order to lower CO_2 emissions, PassivHaus limits energy use for space heating and hot water to 15k Wh/m² per year and total energy use to 120k Wh/m² per year. To achieve this, the walls, roof and floor slab of dwellings must have a U-value not exceeding 0.15 W/m²K and windows not in excess of 0.80 W/m²K. Air infiltration must not exceed 0.6 air changes per hour, which is equivalent to a UK standard of 1 m³/hr/m² – one-tenth of the minimum standard demanded by the 2006 version of

Approved Document L. This is achieved through the use of mechanical ventilation with heat recovery.

The above issues only serve to heighten the already grave concerns over the increasing complexity and future use of cavity wall construction.

3.6 DISCUSSION AND SUMMARY

There are opportunities for significant reductions in the energy embodied in buildings if designers have the opportunity to make informed choices when selecting from alternative materials and those choices can contribute to a general reduction in energy use and resultant environmental damage. The haulage of construction materials within the UK accounts for some 20% of the embodied impact of construction – the more locally that masonry products are sourced, the better. Architects are faced with a massive array of material alternatives when designing a new or refurbished masonry building, and the choices made at this stage are critical due to the impact of production and transport upon the environment.

Although not detailed in its analysis due to the lack of available data, this chapter has shown how manufacturers of some masonry materials are lowering (and are capable of reducing further) the levels of embodied energy associated with their products by investing vast amounts of money in kiln technology.

It is clear that while the cement and lime industries are energy intensive, producing large quantities of CO_2, the cement industry has been able to make substantial technological improvements to production techniques due to the large sums of money being generated by manufacturers. This has vastly improved both energy efficiency and the levels of subsequent CO_2 emissions into the atmosphere. The comparison often drawn between the temperatures required to produce cement and those required to produce hydraulic lime frequently fails to take into account the higher temperatures required for the production of more eminently hydraulic limes, and instead the low burning temperatures for calcium limes are often quoted. There is strong evidence to suggest that claims that one tonne of cement produced equals one tonne of CO_2 are way off the mark. By utilising pre-heaters, pre-calciners and alternative fuels, cement producers are in some instances able to achieve emission levels that amount to less than two-thirds of those commonly quoted by advocates of hydraulic lime binders.

When considering that some masonry units (in particular clay and stone) have shown their ability to last many hundreds of years if used in a suitable manner, the overall effects on the environment of the initial production of the materials is even further reduced over time.

Manufacturers of clay masonry units, like cement manufacturers, are realising that their production methods are energy intensive and are making great strides to achieve efficiency gains and a reduced carbon footprint, thereby dramatically reducing their costs while at the same time improving their corporate profile in the eyes of the public and environmental groups. Waste materials (including those that are toxic) can be used as alternative kiln fuels in the production of cement binders and clay masonry units. The materials are incinerated, reducing their effect on the environment via disposal as landfill. They can also be incorporated into masonry materials in order to lower the demand for fresh non-renewable resources such as clay.

Calcium silicate bricks are a relatively new concept compared with clay masonry units, but offer an environmentally friendly option in terms of the low amount of energy and raw materials used in production. When confronted by the vast array of products on the market, many specifiers might not be aware of the wide variety of colours and textures that calcium silicate manufacturers can now offer.

Modern lightweight concrete masonry systems offer rapid construction times and excellent thermal performance when used for internal blockwork. They require far less energy during manufacture than clay products, with aggregates, and in some cases waste products such as PFA, accounting for up to 80% of the raw material used in their makeup. Although widely promoted, the long-term benefits of a wholesale move towards aircrete buildings would appear to be questionable in terms of the low skill levels required to construct such buildings and, ultimately, these buildings' overall appearance and durability. However, when their overall benefits such as thermal performance, green credentials and ease of use are taken into consideration, it is clear that these materials have a very important part to play in the immediate future of masonry construction, particularly in the construction of internal walls.

The vast majority of stone masonry is extremely pleasing on the eye and lends itself to its surroundings. In terms of sustainability, stone is the obvious choice for facing masonry. Having a durability that has been proved over many thousands of years, it contains little embodied energy if it is sourced locally. However, the high cost of stone and the fact that mass-produced brick and block can be mobilised quickly in today's fast-moving construction climate are major barriers against stone's wider use.

Masonry materials can be reclaimed, used again, and then continue to last for many hundreds of years. It is therefore disappointing that although the masonry sector claims potential for the total reuse of masonry components at the end of a building's useful life, no guidance is currently available from material manufacturers stating how this can best be achieved. It would therefore seem important that urgent consideration be given to the provision and storage of material characteristics and as-built data for future generations of demolition contractors and designers. Only then will fears of litigation due to the specification of reclaimed materials cease to be a barrier to more sustainable masonry construction.

While current insulation options have been considered briefly in this chapter, it seems very obvious that placing more and more insulation products into ever-widening cavities is an unsustainable situation from the point of view of durability and future upgradability. In examining this situation and the government's drive towards zero-carbon buildings, manufacturers of insulation products will need to look very carefully at innovative ways to make their products leaner, while at the same time offering higher levels of performance. As fuel costs continue to rise and reliable sources for gas and oil simultaneously decline, the development of renewable energy sources and heating technology will, in the future, make a difference to the way in which energy is conserved through the use of insulation. It would therefore seem obvious that concentrating on designing durable masonry and at least placing the insulation in an accessible location will give future generations the opportunity to easily upgrade such buildings as and when required.

4 DESIGN CONSIDERATIONS

Buildings are only part of our habitat. Buildings are intimately linked to the local, regional and global environments that are all part of our 'Ecological Niche'. It is the responsibility of our generation to begin to adapt our buildings to ensure that we can stabilise climate change, that we can live without fossil fuels and that we do not unsustainably pollute the environment. Only by so doing can we ensure the survival of our own habitats.

Roaf et al (2004, p. 13)

4.1 INTRODUCTION

Having examined the production and use of new masonry components in Chapter 3, this chapter looks at how masonry buildings can be put together in such a way as to ensure durability, structural flexibility/adaptability, enduring aesthetic appeal and their ability to maintain a suitable internal environment, while at the same time reducing the energy required to do so. The chapter will begin by examining ways in which the subsequent reuse of new masonry materials currently being ploughed into construction projects can not only be made possible, but can become an overall culture that should be entrenched within the demolition and construction processes of the future.

4.2 DESIGN FOR DECONSTRUCTION AND REUSE

Where materials or methods are used that are not referred to by this code, their use is not discouraged, provided that the materials comply with the requirement of the appropriate British Standard and that methods of design and construction are such as to ensure a standard of strength and durability.

BSI (2005a, p. 3)

There is significant potential for reclaimed materials to be used in new masonry construction, but at present numerous barriers exist that restrict their use. In recent years, there has been considerable interest and research into the reuse of construction materials, but little progress has been made in the practical implementation of research and government recommendations (Hobbs and Collins, 1997).

Demolishing one building and disposing of most of the materials, only to build another one (whatever its environmental credentials or lack of them) must be seen as a step back from taking environmental responsibility seriously.

Smith et al (1998, p. 72)

Currently, barriers to the use of reclaimed masonry materials include (CIRIA, 1999a and 1999b):

- The lack of data for reclaimed materials
- The high standard of existing specifications that constrain the use of reclaimed materials unable to meet those standards
- The lack of reclaimed materials available for purchase that meet the specifications
- The perceived liability issues associated with specifying reclaimed materials
- The client's reluctance to include reclaimed materials in a design
- The concern that the client is unwilling to pay for the use of reclaimed materials and will therefore opt for a primary alternative, particularly if the primary material is cheaper (currently, reclaimed materials can be double the cost of those that are new, with genuine Tudor bricks costing three times more than comparable new products)
- The design contract conditions, which may not reward designers for the time or risk they may take

Many building materials arising from demolition can be reclaimed, but there is presently little or no economic incentive for this to happen on a widespread basis. There is clearly a need for a national policy to enforce, rather than encourage, a new culture of reusing perfectly adequate materials generated by demolition, which would otherwise be sent to landfill. The reclamation process has been proved to be successful over many hundreds of years. Masonry units, unlike many other products, do not decompose with time, and it would seem that it is simply too easy to specify new materials and ignore the potential of those that could be reclaimed. Bricks can, however, be compared to a fine wine – they improve with age – and today's mass-produced products have the potential to become tomorrow's highly sought antiques if only thought were to be given to this possibility. As has already been highlighted, MMA (2006) has claimed that masonry construction is totally recyclable/reclaimable at the end of its useful life. Why then do developers continue to rapidly assemble masonry buildings with materials that do not lend themselves to such an approach? For example, how often is a mortar specified on the grounds of assisting reclaimability? The simple answer is not often enough; often should mean at all times.

The advice given by McKay (1944) that cement mortar should be mixed to proportions of 1:3 (cement:aggregate) demonstrates the distinct lack of knowledge of the properties of cement-based mortars in the middle of the twentieth century. Even though the strength of manufactured cement in 1945 would have been roughly half that of today's products due to the vast increases in the strength of the material achieved via improvements in production techniques (Figure 47), such a mix would still provide an eventual compressive strength of around 10 N/mm² today, far stronger than that of many masonry units (Mortar Industry Association, 2004b). It was this type of ignorance of the properties of cement during the early to middle part of the twentieth century that led to widespread damage to much of the UK's historic fabric, repairs having been carried out to soft stonework with cement-based mortars in lieu of the weaker calcium lime mortars used in their original construction. Only during the 1970s did this problem begin to be reversed, leading to the more general vilification of cement from conservation quarters in the late 1990s and early years of the twenty-first century.

Cement mortars were beginning to prove to be far too strong for most masonry applications as early as the 1930s, but even though masonry cements (cements blended with inert void fillers) were developed in the USA in order to alleviate the problem at that time, strong cement mortars have continued to be used to the present day.

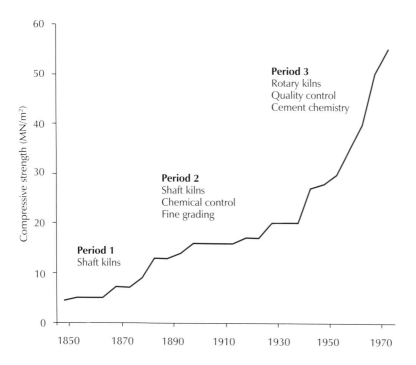

Figure 47: Stages of technological improvement; 28-day compressive strength of Portland cement mortar (1:3) by weight

Whereas masonry cement in the UK has utilised mixtures of Portland cement and crushed or ground stone, in North America the tradition has been to use mixtures of Portland cement and hydrated lime, together with air entrainment. This concept has been adopted in the UK and incorporated into British Standard BS 5628 (Mortar Industry Association, 2004a).

Any one mortar will not usually match all possible requirements. Since the strength of masonry units is highly variable, so too are the requirements of mortars. A cavity wall may be constructed consisting of an inner leaf of blocks with a compressive strength of 10 N/mm^2 and an external leaf of concrete facing bricks with a compressive strength of 50 N/mm^2. Rarely is this taken into account, and the same mortar will generally be specified and used on sites with masonry units of varying strengths. Invariably, this approach of using a 'universal' cement mortar for all applications will occasionally result in excessive mortar strengths. De Vekey (1993) recommends a 'general purpose' mortar for low-rise applications, the mix proportions being 1:1:5½ (cement:calcium lime:sand) with sufficient air-entraining agent to disperse 10–18% of air. Having been incorporated into BS 5628, the US practice of using pre-mixed masonry cements containing calcium limes and air-entraining agents would seem a natural way forward. This concept is being adopted by UK cement manufacturers and is similar in nature to the durable general purpose mix prescribed by BRE.

Typically, a binder-to-sand ratio of 1:3 produces a broadly acceptable mortar mix in terms of the working or plastic properties. However, such cement:sand mortars produce compressive strengths that are far in excess of those required, a 1:3 (cement:sand) mortar achieving a compressive strength of 20 N/mm^2 in 28 days (Mortar Industry Association, 2004b). BRE (1991) recommends compressive strengths of between 2 N/mm^2 and 5 N/mm^2 for low-rise structures, conveniently matching the strength of some standard hydraulic lime mixes. In 28 days, a 1:3 moderately hydraulic (NHL 3.5) mortar mix will achieve a compressive strength of 1.47 N/mm^2, this rising to 3.94 N/mm^2 after six months (Setra Marketing Ltd, 2001).

It is well recognised that mortars with low compressive strengths, such as some hydraulic lime mortars, are ideal when attempting to reclaim masonry units during the demolition of old buildings, although it should be appreciated that eminently hydraulic mixes can achieve strengths similar to some neat cement:sand mixes (up to 11 N/mm^2), making the reclamation of masonry units impractical (BDA, 2001). The stronger the mortar is (or the higher the cementitious content), the greater that the bond strength between the mortar and the masonry unit will be (Mortar Industry Association, 2004c).

Around 10 or so old bricks embody the energy equivalent of a gallon of petrol, and while 3.5 billion bricks are manufactured annually in the UK, 2.5 billion are destroyed. If bricks are reclaimed whole to be reused as bricks, then the energy content is preserved. In Britain around 25 000 tonnes of reclaimable materials are disposed of every day into landfill sites. Around half of this material is from Edwardian and Victorian buildings, and represents a period when materials were made to last.

Woolley et al (1997, p. 59)

A 1:1:6 mortar (cement:calcium lime:sand) or a 1:3½–4 mortar (masonry cement:sand containing an air-entraining agent), like a moderately hydraulic lime mortar, has a low compressive strength of around 3.6 N/mm², the difference being that while a moderately hydraulic lime mortar will take six months to achieve a strength similar to this, the cement mortar can achieve it within 28 days. This quick gain in strength, due to the compounds contained within the cement, is a massive advantage in today's construction climate, where speed is of the essence. As a result of their slower rate of setting, hydraulic lime mortars take longer to achieve resistance to adverse weather for longer periods than Portland cement mortars during laying and curing, be this during cold weather, wet weather or high temperatures (and consequently also require protection from these elements for longer). Hydraulic lime mortars are particularly susceptible to frost.

A compressive strength of 3–4 N/mm², as obtained through the use of a 1:1:6 mix (cement:calcium lime:sand) or a 1:4 mix (premixed masonry cement:sand mix containing calcium lime and admixtures) is ideal for the vast majority of today's cavity constructions due to its ability to gain full strength within 28 days (BSI, 2001a) and would also allow the eventual reclamation of masonry units. One person can clean up approximately 2000 bricks per day if they were originally built in lime mortar (Addis, 2006). However, strong cement mixes continue to be used, which will preclude the future reclamation of masonry units.

As the UK government attempts to move towards more sustainable means of construction, it will no longer be acceptable to simply dispose of whole obsolete buildings to landfill sites. Recycling and reclamation will become the most cost-effective (due to proposed further rises in the landfill tax) and desirable waste-disposal strategy (DETR, 2000). For example, the design and layout of buildings such as supermarkets (Figures 48–50) can become obsolete after a relatively short period, although the masonry units used in their construction are likely to have a lifespan far in excess of the useful life of such buildings.

Because of high transport costs, bricks reclaimed from demolition need to be sourced locally for new projects, generally within a 250-mile radius of the yard to make their use cost effective (Lazarus, 2005). This is usually architecturally beneficial as the more locally sourced the bricks are, the better they will fit into the local vernacular.

Figure 48: Nineteenth century picture of Cockermouth Fire Service in front of Mitchell's cattle auction building, built in 1865

Figure 49: Mitchell's cattle auction building being demolished in 2001 prior to the construction of a new Sainsbury's supermarket. The local planning body had asked that the building be reused due to the architecturally sensitive location of the proposed new building. However, low existing ceiling heights meant that this approach was not feasible.

Figure 50: New Sainsbury's supermarket in Cockermouth. The building is a modern slant on the old auction building, using the sandstone reclaimed during its demolition. Unlike many of today's modern out-of-town stores, this building has been designed to last for many years, in a style that will endure.

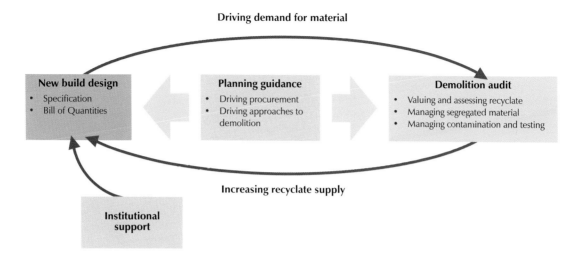

Figure 51: Planning and resource efficiency model for demolition and new build

It is clear that local planning bodies could, if given the powers to do so, ensure that proposed developments are specified on a long- rather than short-term basis. EnviroCentre Ltd (2002) makes this observation and offers a model for the reuse of construction materials, based on the premise that planning guidance is able to drive the process forward (Figure 51).

Pioneering steps have been taken by the London boroughs of Brent, Ealing and Merton, which all require that a sustainability appraisal be submitted with a planning proposal, describing how the building has been designed to reduce its environmental impacts over its lifetime (Roaf, 2004).

Although a clear national planning policy should be enough to drive the reclamation of masonry materials, the fact remains that few technical or financial incentives exist that might encourage the reuse of masonry materials. One measure introduced by the UK government in an attempt to encourage recycling and reuse has been the landfill tax. However, compared with The Netherlands, where landfill tax is levied at a rate of more than double that in the UK (Addis, 2006), it is clear to see why the Dutch are so successful in terms of recycling construction waste, reusing around twice the amount of material resulting from demolition (approximately 90%) than is the case in the UK (Hurley and McGrath, 2001).

The Dutch have also introduced a certification system that provides quality assurance of building products, processes and contractors that use recycled materials. It facilitates the use of reclaimed materials because consumers are able to consult an independent third party in the event of complaints associated with their use (CIRIA, 1999c). Although work on a certification scheme for reclaimed materials has been undertaken by BRE and WRAP (BRE,

2003), little progress has been made in formulating a suitable system in the UK. This situation clearly needs to be resolved before progress can be made and, as was highlighted briefly in the previous chapter, the storage of detailed product and as-built information would be a major move forward in this respect. This would make the demolition audit of buildings far easier for future generations of demolition contractors, who will hopefully become 'reprocessors'.

Overall, it seems clear that due to the lack of available information on existing buildings, the opportunity to reclaim and reuse some of the materials currently being generated by demolition has already been lost. Although the testing and certification of existing reclaimed materials should prove to be relatively straightforward, the overall incentives simply do not exist to facilitate widespread reclamation. The sooner the attitude towards legislative requirements, material production and new construction changes, the quicker this situation will be rectified, making the use of reclaimed materials cheaper and risk free. The supply of reclaimed materials can be unreliable and their costs unpredictable (Addis, 2006). The market will become more reliable as it grows. However, if the durability of today's masonry buildings is maximised, many years will pass before the fruits of this approach are realised. It is important to point out that if this approach were to be followed, the current common approach to masonry unit sizing would need to be frozen in time to ensure the eternal compatibility of all materials.

4.3 DESIGN FOR LONGEVITY

It is the way a building fulfils its function, and the efficiency with which it does it, that gives it value ... The bricks and mortar are simply a means to an end; they do not provide any value in themselves ... Although buildings are usually valued in accounting terms as assets, their value is generally not inherent; it is derived from their function ... If the need that a building was built to fulfill disappears, then the building's value will derive from its ability to be adapted to an alternative use.

Roaf (2004, p. 19)

4.3.1 Durability and adaptability

While masonry buildings constructed with solid walls have proved beyond doubt their ability to last for many centuries, the jury is still out on those constructed more recently with cavity walls. Cities such as London, Bath (Figure 52) and Edinburgh contain many solid walled buildings that are as popular today as they were 200 or even 300 years ago, and having paid the initial investment in environmental costs, they can now be periodically upgraded at relatively low cost by successive users. Conversely, it has been shown that cavity wall construction, having been popular for only 60 years or so, needs

Figure 52: The Circus, Bath

regular attention in order to avoid the development of excessive wall tie corrosion (Figures 53 and 54), which can become the cause of major structural defects and water ingress (Harrison and de Vekey, 1998; de Vekey, 1993; Brand, 1994).

Figure 53: Horizontal cracking of bed joints due to wall tie failure

Figure 54: Corroded wall ties

The vast majority of wall tie failures occur when galvanised ties are used. The more recent developments of stainless steel, duplex-coated zinc, copper alloy, polypropylene and PVC-coated mild steel offer improved durability, but they are all susceptible to far quicker deterioration rates than the masonry around them, although exact timescales of likely deterioration are unknown due to the relatively short time that these products have been in use (Harrison and de Vekey, 1998). In addition to the problems of wall tie deterioration, concerns have recently been raised in relation to buildability and workmanship issues around cavity wall construction, including those from Baiche et al (2007), Brand (1994), de Vekey et al (1989), Harrison and de Vekey (1998), Hazael (1993), Howell (1995), Mason (1992), McNeilly (1993), Roberts (1998), Sutherland (1993) and Thomas (1989), as follows:

- Inadequately protected or missing ties
- Ties sloping the wrong way
- Inadequately embedded ties
- Cracking and instability due to corrosion or absence of wall ties
- Mortar on ties
- Reduced weather tightness due to induced cracking
- Cavity trays omitted or poorly detailed, leading to damp penetrating to the inner leaf
- Stop ends of cavity trays omitted
- Rotation of concrete boot lintels
- Weep holes to cavity trays absent or blocked
- Cavities too narrow
- Cavities bridged
- Cavity fill deterioration
- Corrosion of inadequately protected or damaged steel lintels
- Ties providing no shear connection between the two leaves
- Poor resistance to fire in comparison with solid construction
- Missing or misplaced damp-proof courses
- Difficulties associated with movement-control joints (a problem not previously experienced with solid wall construction due to its greater mass)
- Increasing complexity due to changes to Part L of the Building Regulations

In light of such concerns, recent research has been carried out into ways of avoiding cavity construction. Traditional Plus, developed by CERAM, is a single-skin masonry system designed for low-rise construction (up to two storeys). This form of construction relies on the use of a large cellular clay unit (290 x 140 x 65 mm) in order to obtain enhanced stability during construction. The single skin of masonry is lined internally with a waterproof insulant, to which is fixed the internal finish (Figure 55). The system fully complies with UK Building Regulations, although uptake has been extremely slow since its development.

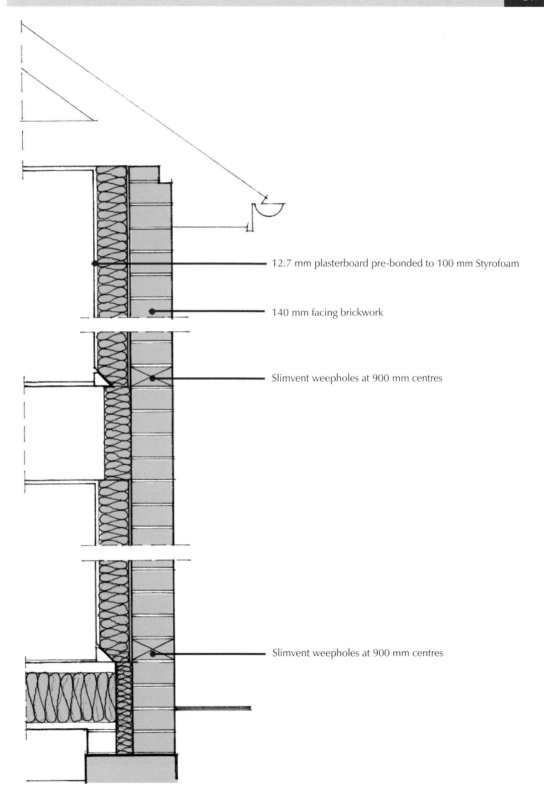

Figure 55: A cross-section through the Traditional Plus system

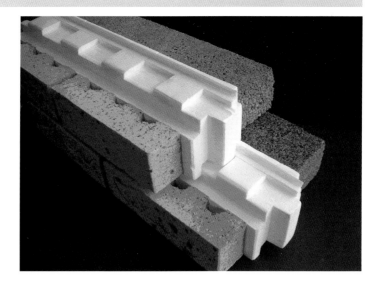

Figure 56: Two courses of the composite masonry units developed by Oxford Brookes University and sponsored by Hanson plc

Hanson plc proposed a project to Oxford Brookes University in 2007 to explore the development of a composite masonry unit (Baiche *et al*, 2007). This resulted in the development of units comprising two brick slips, a shallow concrete block and a polystyrene core (Figure 56).

However, as well as the problems of water ingress, condensation, inability to meet the requirements of Part L of the Building Regulations, poor buildability and low flexural strength encountered during this project, grave doubts would have to be expressed from a sustainability point of view about the durability and eventual reclaimability of such a product.

The fact remains that much of Western Europe still utilises the more traditional solid wall method of construction with great success. Although the climate experienced in the countries in which this form of construction is used – ie Spain, France, Germany (Roberts, 1993) – might be considered as more temperate than that of the UK, a simple question must be asked: Would twenty-first century technology allow the construction of watertight solid walled masonry buildings?

One simple fact has been observed by the author during over 25 years spent working in the construction industry: extremely old buildings with solid walls have continued to be adapted easily and have had the capacity to meet whatever challenges the Building Regulations have thrown at them. Each year, hundreds of proposals are submitted to planning and building control bodies for the conversion of such structures as barns, mills and factories. During their lifespan, these buildings have presented few hidden dangers such as corroding or missing ties or degrading insulation – what you see is pretty much what you get. One such project in which the author has had involvement is the conversion of an old iron ore pit building into a dwelling on Winscales Moor near Egremont, Cumbria (Figures 57 and 58). This building consists of 350 mm solid walls

Figure 57: Old pit building on Winscales Moor prior to works starting

Figure 58: Pit building being given new life as a dwelling

in clay brickwork and sits on a moor overlooking the Solway Firth, one of the most exposed locations in England. Despite this, the walls, which were extremely well built, showed no signs of water ingress prior to works starting. As part of the renovation works, the walls were lined internally with a vapour barrier and insulation and rendered externally, and will not present a maintenance problem to the occupiers for many years to come. When the insulation needs to be upgraded, this can be done without having any structural implications.

The fact is that although it has proved to be extremely durable, the main reason that solid wall construction went wildly out of fashion around 60 years ago was that it sometimes failed miserably on the grounds of water penetration. Although British Standard BS 5628-3 (BSI, 2001a) suggests that an unrendered solid wall with a thickness of at least 440 mm should only be used in sheltered locations (Table 9), this does not seem to take into account the many products now available to designers. After all, CERAM has designed an unrendered walling system that is only 140 mm thick and is watertight. Cellular clay blocks such as those manufactured by Ziegel have proved to be impenetrable. It is also worth noting that although more masonry is required for solid walls than with cavity construction, the effort and costs involved are actually comparable due to the array of parts such as ties, damp-proof courses and trays required in cavity wall construction (Roberts, 1993).

Table 9: **Single-leaf masonry: recommended thickness of masonry for different types of construction and categories of exposure**

Type of masonry	Minimum constructional thickness (mm)*	Maximum recommended exposure zone for each construction (1 = sheltered; 2 = moderate; 3 = severe; 4 = very severe)			
		Unrendered	Rendered in accordance with BS 5262	Externally insulated	Impervious cladding
Clay or calcium silicate	90	Not recommended	1	3	4
	215	1	2	3	4
	328	1	3	3	4
	440	2	3	3	4
Dense concrete	90	Not recommended	1	3	4
	215	1	2	3	4
	250	1	3	3	4
	328	1	3	3	4
	440	2	3	3	4
Lightweight concrete	90	Not recommended	1	3	4
	190	Not recommended	2	3	4
	215	Not recommended for blocks with open surface texture	3	3	4
	328	1	3	3	4
	440	2	3	3	4

* Based on work sizes and excluding render thickness.

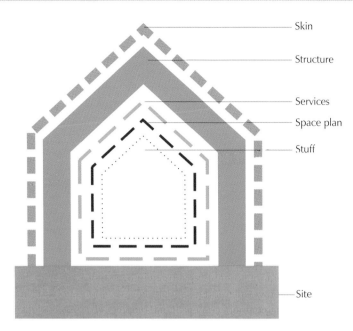

Figure 59: Brand's 'Shearing layers of change', with each layer requiring replacement at intervals from three years (space plan) up to 300 years (structure)

Brand (1994) describes the concept of 'building layers' (Figure 59) whereby the 'six S's' (skin, structure, services, space plan, stuff and site) need to be replaced at intervals reasonable to their ability to endure the changing patterns of society. He postulates that the structure of a building should be able to last for up to 300 years, which is not unreasonable for solid masonry construction. In more recent years, a seventh 'conditioning' layer seems to have evolved within buildings built with solid masonry; this is an internal lining of boarding or studwork incorporating insulation to bring these buildings up to an acceptable standard in terms of energy efficiency. As the insulants improve over time, so this layer continues to evolve, and while the masonry structure has been built to stand the test of time, this layer, which can be made up of renewable/recyclable materials, can be renewed or upgraded at required intervals. This makes these buildings not only durable, but extremely adaptable.

The economic and environmental benefits of renovation in lieu of demolition and new construction are well recognised (Highfield, 2000). If buildings are to be built today with future requirements in mind, designers need to start designing what Brand describes as 'scenario buffered buildings'. This means predicting the likely future requirements of buildings instead of only designing for the speed of construction and lowest cost, which seem to have become the main considerations of today's clients, construction professionals, the industry in general and the UK government.

Another advantage that solid wall construction can claim over cavity walls is the ability to accommodate movement, thereby reducing the need for expansion joints. Most materials will expand or contract due to changes to the thermal environment around them and in their own moisture content, including masonry units. It is possible to take full account of movement in a successful design, but there are many designers who are not completely aware of the importance of the subject, failing to appreciate that movement is a parameter that needs to be considered at the initial design stage.

Claims that the use of hydraulic lime mortar can negate the need for expansion joints, regardless of the type of wall construction, are common. Large new buildings such as the RSPB headquarters in Bedfordshire and Haberdashers' Hall in London have been built with hydraulic lime mortars; no expansion joints have been incorporated into the masonry of either building.

Research on clay brickwork carried out between 1962 and 1983 indicates that moisture movement can take place over a period of up to 20 years after construction, with the amount of movement taking place at and beyond this period being of no practical significance. The magnitude of the movement during the 21-year period of monitoring was up to 1 mm/m run of wall (Morton, 1986).

No reliable data exist that can substantiate the claim that the use of hydraulic lime mortars removes the need to incorporate expansion joints. British Standard BS 5628-3 recommends joints at centres as low as 6 m (concrete masonry) and as high as 15 m (clay masonry with expert advice), depending upon the expansion coefficient of the masonry units being used. This rarely seems to be taken into account when masonry buildings are designed.

BDA (2001, p. 4) holds the view that far from being solely attributable to the properties of lime mortar, historical examples of buildings built with no expansion joints owe much to the mass of solid brickwork, which highly stressed cavity construction does not possess:

No authoritative guidance is available on this aspect of design [the avoidance of movement joints]. Historical examples of masonry with lime mortar and no movement joints are always thick heavyweight brickwork. Great weight imparts restraint and thick walls are able to absorb more heat than thin ones and so they react less to thermal change. The mass of masonry is more significant in limiting movement than the lime mortar.

BRE (1991) states that mortars (including cement-based mortars) within its recommended compressive strength range of 2–5 N/mm^2 will accommodate small movements, with any cracking in the masonry tending to be distributed as hairline cracks in the joints where they are unobtrusive and do not prejudice stability.

4.3.2 Consideration of whole life costs

The UK government has championed the use of MMC, claiming that traditional methods of construction are too slow to cope with the rapidly growing demand for new housing (BURA, 2005), and has even invited designs for the '£60 000 house'. However, rapid construction techniques, while serving a short-term accommodation problem, do not normally make for long-term solutions in terms of buildings that do not require early replacement – as proved in the 1960s and 1970s (Highfield, 2000). This approach would appear to be a false economy, and it is worth pointing out that masonry buildings are far cheaper to construct than those built using MMC (BURA, 2005).

British Standard BS ISO 15686-1 (*Buildings and constructed assets – Service life planning – General principles*) (BSI, 2000, p. 28) defines WLC as an:

… economic assessment considering all agreed projected significant and relevant cost flows over a period of analysis expressed in monetary value. The projected costs are those needed to achieve defined levels of performance, including reliability, safety and availability.

It is a process that aims to look at every cost incurred in respect of a facility or product from inception to disposal, and its objective is to make investment decisions with a full understanding of the cost consequences of different initial options (Bourke *et al*, 2005). Although there seems to be a growing recognition for the need for building designers to consider a whole life and sustainable approach to construction, some clients are still reluctant to do so. To ensure a more sustainable approach to masonry construction, there is an urgent need to develop and promote the use of WLC to enable the achievement of a more durable built environment. Clients need to be made aware that construction methods that are initially the quickest and cheapest will eventually result in the least satisfactory and most costly structures through the need for early replacement. If the life of buildings is short, the decisions made during their conception will have a marked impact upon the environment well into the future when energy, land and raw materials may be restricted. Therefore, it is in the interests of both clients and design professionals to prolong the lifespan of buildings.

4.3.3 Masonry quality and craftsmanship: do they still exist?
Bricklayers competing at lowest price on a chaotic housing site; walls programmed to be built mid-winter; bricks barely inside British Standard, if not actual seconds; a design where the walls have to be cut around non-brick-sized British Standard windows. This is no way to build houses that will have much durability, or beauty … Faster and cheaper is not always more efficient. Slower site work for higher pay is an investment in time worth making if we want to raise 'building' to 'architecture'. It is an investment in civilisation more immediately valuable than all the rhetoric about 'sustainability'.

Abley (2007, p. 11)

Advocates of MMC suggest that the required numbers of skilled tradespeople simply are not available any more:

Figure 60: The rebuilt west elevation of St Pancras Station in London, which was voted the 'supreme winner' at the Brick Awards 2006. The fine clay facing brickwork was built in lime mortar with a joint thickness of 5 mm, requiring superb skill levels, which were relatively easily attained by the bricklaying contractor, Irvine-Whitlock, who kindly provided this picture.

The scrapping of apprenticeship schemes, in tandem with the general lack of interest in taking up skilled building work as a career, has led to a reduction in the number of skilled workers available to contractors.

BURA (2005, p. 17)

However, there is also a view that MMC is contributing to this perceived problem:

I believe the increase in MMC will contribute to the further dumbing-down of skilled tradespeople, which will be a major problem for repair and maintenance work in the future.

Jones (2007, pp. 33–34)

There is no doubt that for projects where high skill levels are required, the necessary calibre of tradesperson is available; the high standard of the nominees for the Brick Awards and Natural Stone Awards each year certainly bear testament to that (Figures 60 and 61). It could well be the case that in attempting to follow the lead of the UK government, the masonry sector is, in some respects, shooting itself in the foot by putting forward its own 'masonry MMC' (ie prefabricated walls built with high strength 'glue' mortars) rather than improving on what it does best and arguing the case for more traditional forms of masonry, based on previous performance.

Perhaps, as Abley (2007) suggests, the industry should be looking to recognise and pay for high skill levels, as used to be the case historically when masonry was regarded as a profession, rather than pay bonuses for the amount of work done per day, as would normally be paid to unskilled factory workers. The fact is that there are currently around 120 000 bricklayers in the UK, and if each laid 500 bricks per day, the annual output of the brick industry would be laid in 10 weeks (BDA, 2007; Irvine-Whitlock, 2007). Instead of looking

Figure 61: The Jerwood Centre in Grasmere, Cumbria, which was a highly commended entry in the 'slate' category of the Natural Stone Awards 2006. Built by K R Venning Slaters Ltd, the external leaf of the building is made up of random Burlington slate walling. The building (which houses the Wordsworth Trust's Romantic Movement manuscripts, books and paintings) has won critical acclaim from RIBA and the Civic Trust, and although modern in nature, blends spectacularly well into its traditional Lakeland environment.

for quick, short-term solutions and profit, clients and designers need to start to consider the quality of the end product.

Designers can assist in the achievement of quality workmanship by considering 'buildability' issues, including the standardisation of building measurements to ensure that brick/block bonding can be easily maintained, which also helps to dramatically reduce waste through the excessive cutting of masonry units (Howell, 1995). Meanwhile, clients need to start considering how procuring high-quality, sustainable buildings can enhance their reputation with stakeholders including the public, pressure groups and investors (CIOB, 2002). The quality of some of our ageing built environment is often enthused over and stark comparisons are made with the construction industry of today. However, what needs to be remembered is that historically, it is only the well-built buildings that have survived; cowboy builders and poor paths of procurement have always existed (Briggs, 1925).

Another criticism aimed at more traditional methods of construction is that poor site conditions do not permit the development of a quality construction product, and it is suggested that prefabricated components are the way forward (BURA, 2005; Hogg et al, 2002; Hendry and Khalaf, 2001). This is not necessarily the case, however, and is again a case of the construction industry becoming complacent and lazy.

A prime example exists on the author's doorstep, in Cumbria. Whereas some volume house builders' sites are poorly organised and give little consideration to maintaining a suitable production environment, one developer stands out as being head and shoulders above the competition in this respect. High Grange Developments, based in Whitehaven, is renowned for the quality of its finished product and takes great care in considering how this should be achieved. While most contractors leave themselves open to defects and lost time due to the often inclement British weather, High Grange Developments takes the time to 'cocoon' its products from the elements, using large tented structures (Figure 62). This approach has proved to be extremely cost effective and the finished products (which are usually award winning) show that, in terms of quality, the effort is well worthwhile (Figure 63).

Figure 62: Shelter over a house being constructed at Mariners Way, Whitehaven

Figure 63: The finished product – one of the houses built by High Grange Developments

The quality of modern manufactured masonry units can usually be guaranteed and can at least be quality-controlled via checks upon delivery to site. However, it is worth pointing out that if natural stone is used as the external facing material, decay can result from a number of causes arising from poor quality control and workmanship. For example, incorrect bedding (where the stone is not built with its natural bed perpendicular to vertical load pressure) can result in serious lamination of the stone; dissimilar materials and variability (eg mixing sandstone with limestone) can result in decay from chemical transfer from one type of stone to another.

The major factor often overlooked in the pursuit of high-quality masonry is the production of mortar. Mortar is seen as a means to an end rather than a product whose quality is as essential as that of other building components. Specifiers pay it scant regard and operatives mixing on site are more concerned with keeping bricklayers happy; the finished product is usually made workable to the point of variance to the actual specification, if indeed a specification is provided.

Many of the claims that cement mortars are too strong for general masonry applications appear to stem from the conservation sector and cases where strong cement mortars have been used to restore or repoint old buildings built with soft stones. Wider experience seems to suggest that when site-batched mortars are used on new-build projects, the mixes are, more often than not, far too weak. The article 'Get it right: masonry walls' (Cuffe, *Building*, 31 October 2003) highlights the fact that problems with mortar to external masonry walls come high on the list of recorded defects, this according to home insurance claims and industry statistics. In the article, Nick Cuffe, technical manager at Zurich Insurance Building Guarantee, states that excessive weathering and cracking of some masonry was being attributed to weak mortar mixes and that:

The mix should take account of the exposure of the development, the type of masonry and the positioning of the masonry on the dwelling. It is vital, if mixing on site, that there is adequate supervision to ensure the specification is adhered to.

These thoughts are echoed by Harrison (1993, p. 4):

The bricklayer may be unaware of, or unable to avoid, building-in latent defects if the properties of the mortar with which he is supplied are not suited to the size, weight, suction and moisture movement of the masonry units being laid.

Long (1989) details a survey of 106 companies of various sizes, carried out at the end of 1982 by BEC's Technical Advisory Service, to find out what was happening on site at that time in relation to the batching of mortar. Ready-mixed lime:sand was rarely used. In just about 60% of cases, a shovel was used to batch 'dry' materials, this being very inaccurate as a shovelful of damp sand is much greater in volume than a shovelful of dry cement. Trials carried out by the Technical Advisory Service and two research organisations showed that when mixing mortars in free-fall drum mixers in typical site conditions, it was easy to carefully batch 1:4 into the machine and take 1:20 or worse from it, the remainder of the cement being left behind sticking to the drum.

In truth, there are few recorded instances of masonry defects caused by cement mortars that were too strong, but conversely, there are many claims of defects resulting from cement mortars that were too weak and of poor site batching:

Although it is possible to find situations where excessive binder has been used, the vast majority of mortar with incorrect mix proportions has insufficient binder and this is particularly the case where materials are mixed on the building site.

Mortar Industry Association (2004b, p. 16)

Particularly in the site manufacture of the mortar there is considerable scope for improvement, both in improving the consistency of properties of the mortar, and, most importantly, in terms of the durability of the finished product.

Tutt (1990, p. 1)

The operative making the mortar has two overriding concerns: to provide mortar at the rate required by the brick/block-layers, and to produce a workable mortar. Given the materials at his disposal he will achieve these ends in whatever way he can, irrespective of the Code or Specification. Batching will be done with a shovel, provided the Architect/Engineer/Clerk of Works is not looking; water will be taken from an oil drum with a safety helmet and the mixing time will be dictated by the demands of the bricklayer.

Tutt (1990, p. 1)

There have been a number of very costly mortar failures arising from material that has been poorly mixed with low cement content.

Housebuilder (online)

The choice of aggregate is especially essential to the formation of a successful hydraulic lime mortar mix. The best mortar sands generally have a void ratio of around 33–35% (Historic Scotland, 1998), which means that they will require the addition of around one-third of their dry compacted volume in binder paste in order to fill all the voids. This is the basis of the traditional 1:3 binder:aggregate recipes for modern mortars.

Given the poor record with regard to the site production of more familiar and user-friendly cement mortars, concerns over the mixing of hydraulic lime mixes by operatives unfamiliar with their properties would seem justifiable, especially as different mix strengths may be required for different areas of sites, depending upon exposure to the elements. According to the Foresight Lime Research Team, lime mortars mixed in conventional drum mixers are prone to balling, although a batching sequence is described within its work that claims to reduce this. The team also states that for larger projects a rollpan or paddle mixer is preferable. Hydraulic lime mixes need to be given at least 15 minutes to combine. For cement mortars, British Standard BS 5628-3 (BSI, 2001a, p. 95) simply states:

In general, a machine mixing time of 3 min to 5 min after all the constituents have been added should be sufficient.

As detailed earlier, because of their slow rate of set, hydraulic lime mortars take far longer than cement-based mixes to achieve resistance to adverse weather, particularly frost. In terms of durability, all masonry, whether built with cement or lime mortars, is particularly susceptible to freeze/thaw cycles during winter months. The Foresight Lime Research Team (2003, p. 30) recommends the following onerous measures in such conditions:

The structure should be protected with damp hessian to preserve the moisture and with sufficient cover, using bubble wrap or insulating material, to protect the structure itself and the mortar against frost. Added precautions may be necessary, possibly including some form of heating, to protect finished work during prolonged periods of freezing temperatures.

The research of the Foresight Lime Research Team has found that when set, moderately and eminently hydraulic lime mortars exhibit excellent freeze/thaw resistance, as do masonry cement and other cement/lime mixes.

High-quality mortars are essential in the pursuit of durable high-quality buildings. In this respect, accurate specification of mortar mixes and poor-quality hand batching on site is a serious issue that designers and the wider masonry sector need to address. Paint is not mixed on site. Sealants are not mixed on site. Why continue to batch mortars on site with a shovel when high-quality factory-produced mixes are available?

4.3.4 The quest for enduring aesthetic appeal: the new vernacular

Before we had such a cornucopia of materials able to be delivered to everyone's doorstep and building site, almost everything was built from the same local materials. This is why we have vernaculars. If, however, residents of Bath had been able to use limestone from Cumbria or granite from Scotland, perhaps their city would be a bit different today. Perhaps it would look like every other city?

Smith *et al* (1998, p. 81)

Many of the buildings constructed in the last 60 years or so (supermarkets again being a prime example) can become obsolete after a relatively short period due to their quirky 'of the moment' designs, although the masonry units used in their construction are likely to have a lifespan far in excess of the useful life of such buildings. It would appear that designers and the UK planning system have done little to reverse this trend, as billions of perfectly good masonry units continue to be sent to landfill.

Vernacular buildings by their very nature reflect the materials of their immediate locality. One of the problems with today's masonry buildings is that many have become bland and predictable, and do not reflect the geology of the localities in which they are built. Masonry units are now mass produced, and instead of having buildings that fit into the local landscape and the vernacular, one can now drive around the country and view buildings that are indistinguishable in this sense (Blackman, 2007).

The stretcher bond pattern, synonymous with cavity wall construction, offers little aesthetically in comparison with the rich diversity of the English and Flemish bonds used to build our most precious brick structures (Figures 64–66), and while it is possible to use snap headers (or half bricks) to replicate more aesthetically pleasing brick bonds in cavity construction, such techniques are very rarely used (de Vekey, 1993; Harrison, 1989).

Figure 64: English bond   *Figure 65:* Flemish bond   *Figure 66:* Stretcher bond

Figure 67: Cavity wall built with sandstone facing blocks that have been sized to course with the concrete blockwork internal leaf

Figure 68: An extension built with rendered cavity blockwork, thereby avoiding the difficulties of coursing random rubble stonework and internal blockwork

It is also true that we now rarely see buildings built with local stones (eg the limestone of Bath or the slate of Cumbria), which tourists from all over the globe flock to see. As well as cost restrictions against the current use of natural stone as an external finish, cavity wall construction makes its use all the more difficult because of the need to course level with an internal leaf of masonry, this, more often than not, being made up with lightweight concrete blockwork. Ashlar, with its level coursing, can be made compatible with the coursing requirements of modern blockwork (Figure 67). Although possible, it is immensely difficult to use aesthetically pleasing uncoursed random rubble stonework and achieve the desired coursing requirements of blockwork (Figure 68).

The masonry buildings that we build today should last well into the future, but if we do not change our ways, prematurely obsolete buildings will result in rapid degradation of our environment and further endanger our society (Oxley, 2003).

While no one is advocating a return to Victorian or Georgian values, it is clear that we can learn much from our ancestors. A new vernacular should seek to use the best of what we have around us by taking a mix of traditional and modern construction techniques and minimising the environmental impacts of new buildings by giving them enduring appeal (Addis, 2006; Roaf et al, 2004; Smith et al, 1998; Bowyer, 1993; Oxley, 2003). After all, well over 200 years have passed since their construction, but people will still clamour to purchase a Georgian building. Our forefathers must have done something right!

4.3.5 Design for climate change

If one were to be washed up, Robinson Crusoe-like, on one of those mythical 'desert islands' so beloved of story tellers and reality TV, then the first determinant of shelter – leaving aside for the moment any notion of 'architecture' – would undoubtedly be climate.

Best and de Valence (2002, p. 21)

Masonry buildings provide excellent levels of thermal mass, an extremely important property to possess in the light of predicted climate change. However, it is alarming that in terms of the requirements for new housing, the Code for Sustainable Homes concentrates fully on preventing climate change and gives scant regard to coping with it; thermal mass is given very little consideration.

Both cavity and solid wall constructions offer excellent thermal mass, but the optimum solution is to build solid walls and insulate them externally, thereby keeping the cold surface to the inside while creating a 'thermal store' from the outside (Concrete Centre, 2006; Beggs, 2002). Systems are available on the market whereby buildings can be insulated and then rendered externally, and although their lifespan is thought to be around 30 years (Permarock, 2006), it should be borne in mind that in such instances the insulants remain accessible and easy to replace when necessary. Brick and stone effect finishes are available, but the systems preclude the option of an authentic faced masonry finish.

The use of such systems allows solid walls to be built in severely exposed locations and a thickness of 80 mm of insulant will give a one-brick-thick wall a U-value of 0.25 W/m^2K, well within the current requirements of Part L of the Building Regulations. If the insulation is fixed to the inside of a solid wall, a vapour barrier (and in some instances ventilation) should be used to reduce the risk of interstitial condensation affecting internal finishes (Figure 69). Solid wall construction also has an inherent ability to provide far greater levels of airtightness when compared with cavity construction (Aircrete Bureau, 2005), thereby making the achievement of an energy-efficient internal environment far easier.

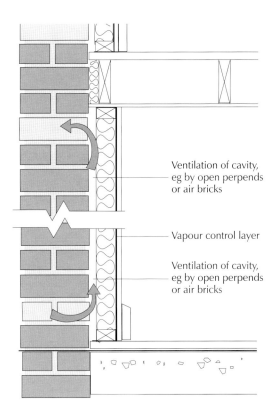

Figure 69: Wall insulated on the inside

As the government presses for a reduction in energy use through conservation measures enforced through the Building Regulations, planning legislation does little to encourage the use of designs that might help to create a suitable indoor environment while at the same time utilising our predicted warmer future climate to do so (Town and Country Planning Association, 2007; Roaf, 2004). Housing developers continue to cram as many detached or semi-detached units onto a plot of land as they can, paying scant regard to how the dwellings will react to the environment around them.

Also ignored is the fact that, due to the heat shared through their party walls, terraced properties are the most energy-efficient form of housing (Edwards, 1999) and have also been the fastest-rising property type in terms of house sales since the mid-1990s (BBC News (online) (c)). Terraces of town houses have been the bedrock of sustainable communities for generations.

Figure 70: The BedZED development in Sutton

A well-publicised but ingenious terraced development is Beddington Zero Energy Development (BedZED) in the Borough of Sutton in London (Energy Efficiency Best Practice Programme, 2002). With south-facing terraces, the mix of one- and two-bedroom flats, maisonettes and town houses are integrated so that one building does not steal sunlight from its neighbours. The scheme is highly optimised and approaches the highest density for mixed use capable of benefiting from useful amounts of passive solar gain, using the heat of the sun (gained from the glazed elevation and stored in the mass of the masonry) to help condition the units and thereby reduce the need for conventional heating (Figure 70).

Roaf *et al* (2004) describe a development on a 64 ha site in Budapest, Hungary, where an original plan for 380 houses eventually ended up being 300 in order to orientate them to the south to make the most of passive solar design (Figure 71), resulting, as in the case of BedZED, in a highly successful project.

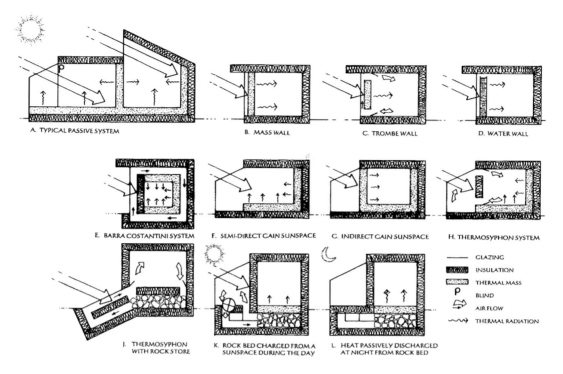

Figure 71: Passive solar systems: (A) typical passive solar system; (B) mass wall system; (C) Trombe wall system; (D) water wall system; (E) Barra Constantini system; (F) semi-direct gain sunspace; (G) indirect gain sunspace; (H) thermosyphon system; (J) thermosyphon system with rock bed; (K) underfloor rock bed actively charged from a sunspace during the day; (L) underfloor rock bed passively discharged by radiation and convection at night.

4.4 DISCUSSION AND SUMMARY

Although there is currently significant potential for reclaimed materials to be used in new masonry construction, even greater potential exists to ensure their future use by way of using weak masonry mortars that facilitate further reuse, along with the accurate recording and storage of as-built information. Masonry cement and cement/lime mixes can match the low compressive strength of hydraulic lime mortars, thereby also facilitating the reclamation of masonry units, the major concern here being that unless the industry moves towards the widespread use of factory-produced mixes in lieu of batching on site by shovel, major defects and a resultant harmful effect on masonry's reputation will ensue.

Massive databases currently exist for the sale of property and cars, so why not create an as-built database for future deconstruction? Availability of local materials, potential for recycling and impact of production should be considered at all stages of design.

In terms of the design of masonry buildings for longevity, one thing seems clear: the continued use of cavity wall construction needs to be seriously questioned, and while innovative developments such as CERAM's Traditional Plus are interesting, the time has surely come to investigate the reintroduction of more conventional means of obtaining durable and adaptable solid wall buildings. Mile upon mile of stretcher bond around the UK offers little visual variation, particularly where large uninterrupted areas of brickwork are necessary. It cannot be beyond the means of the twenty-first century UK masonry sector to come up with a suitable solution, perhaps by mixing traditional facing brickwork or stonework methods (which are much loved and valued by British society) with innovative block and insulant technology.

It would be a great disappointment if the masonry sector were to cave in to the ridiculously short-sighted approach of the UK government, and in doing so begin to offer solutions on the basis of rapid, cheap construction methods. Although innovation cannot be ignored, as an overall approach this is not in the spirit of the highly skilled masonry craft that has built up a reputation over many centuries. Despite reports to the contrary, the UK construction industry still possesses the highest levels of skill required to execute exquisite facing brickwork when it is required. Masonry buildings are far cheaper than MMC, and so it would seem to make sense to invest in skills, improved site practices and masonry buildings of a higher quality than those currently being built. Surely if people are prepared to pay more for MMC, they would rather pay a comparable price for a product that is proved to be far more superior and durable?

Domestic and commercial masonry developments continue to optimise site density and are rarely planned to take account of their orientation, when by doing so they could maximise their potential to naturally create a suitable indoor environment and minimise energy use. In terms of new housing, developers continue to juxtapose predictable detached and semi-detached property types, squeezing them onto tight belts of land without giving thought to optimising shading or passive solar gain. Despite this, terraced housing has proved to be the most popular choice for house buyers since the mid-1990s and has the potential to maximise both density and climatic orientation, while at the same time being far more energy efficient than other forms of housing.

It is important to design to reduce future climate change, but it is equally (if not more) important to design masonry buildings that will be capable of coping with climate change; worryingly, the Code for Sustainable Homes does not appear to recognise this. While a change to planning legislation could do much to increase the quality of future development in this respect, it would seem ludicrous to continue to let planning and building control bodies work in isolation from each other when each has so much knowledge to share in terms of design for both energy conservation and climate change.

Door head detail to restoration and extension of a historic farm complex in the Cotswolds Designed by Robert Adam of Robert Adam Architects and built by Alfred Groves & Sons in Syreford Cream Cotswold stone, the project was highly commended in the 'new build' category of the Natural Stone Awards 2008

5 THE INDUSTRY'S VIEWS

A major advantage of the interview is its adaptability. A skilful interviewer can follow up ideas, probe responses and investigate motives and feelings, which the questionnaire can never do ... Questionnaire responses have to be taken at face value, but a response in an interview can be developed and clarified.

Bell (2002 p. 135)

5.1 INTRODUCTION

The previous three chapters are the result of the examination and analysis of hundreds of texts relevant to masonry construction, which in turn covered hundreds of years of development of masonry form and function. Never in the history of the built environment in the UK has the pressure to improve the performance, form and function of buildings been as great as now, in the early part of the twenty-first century.

Worldwide demands for substantial reductions in CO_2 emissions and the use of raw materials are being recognised through increasingly onerous legislative requirements being introduced by the UK government. The masonry sector, quite rightly, often advertises the fact that although high levels of energy are used to manufacture some masonry units, they can last for hundreds of years and can eventually be reusable. However, the findings of the previous chapters suggest that masonry buildings are very rarely constructed in a way that takes maximum advantage of the qualities of these materials or of how the buildings can best interact with their environment.

It was therefore vital, having examined all available historical literature on the subject matter, to gain a 'snapshot' of the current views of industry through people with the requisite expertise to enable them to comment on matters that had not previously been considered in detail through any other study or publication. Only following an investigation of up-to-date industry opinion could an analysis of the basis for sustainable masonry construction be considered to be rounded and complete.

Consequently, this chapter analyses the content of semi-structured interviews that were carried out by the author in conjunction with the information contained in the previous three chapters. The following pages contain details of the interviewees' thoughts and feelings in relation to sustainable masonry construction for low- to medium-rise buildings.

A variety of techniques could have been used to collect views relating to the subject matter. Questionnaires and structured interviews were inappropriate as investigations can be inadvertently narrowed and biased by the author's field of knowledge and lack the flexibility to probe issues not anticipated. One-to-one interviews were carried out because it was the best way to acquire the information, as the potential interviewees with the required expertise were few in number. The semi-structured nature of the interviews enabled a degree of comparability between the questions asked but allowed room for each interviewee to go into greater detail on issues of particular relevance to him/her.

A significant potential weakness with the interview results was that they were only relevant to the expertise contained within the field of masonry construction and not the construction community and society in general. Therefore, to address this, triangulation of the interview information with the wider information contained in the previous chapters was used. Wider findings have been mapped against the interview results to confirm these as valid issues common to the industry and wider society and not merely the masonry sector.

A total of 12 interviews were undertaken during July and August 2007, each interviewee having experience relevant to the subject matter to which each question relates. The knowledge and experience in masonry and sustainable construction possessed by the chosen interviewees is among the best in the UK, if not the world. Each interview was in depth (some lasting well over one and a half hours), and while being representative of each interviewee's responses, the information presented in the following pages is only a small sample of the data collected through the 12 semi-structured interviews. The list of interviewees is shown in Table 10. The interview data follow the same order as that set out in Chapters 2–4.

5 THE INDUSTRY'S VIEWS

Table 10: **Interviewees**

Interviewee	Background
Ian Abley	Architect, author and head of audacity, a company that advocates developing the man-made environment. Ian has been sponsored by the Modern Masonry Alliance on a four-year engineering doctorate at Loughborough University, the theme of the doctorate being 'Better Built in Masonry'.
Michael Burdett	Programme leader of trowel occupations at York College and UK training manager for bricklaying.
Kevin Calpin	Trained as a stonemason at York Minster and is programme leader of stonemasonry at York College. Received the 2006 President's Award from the City & Guilds Institute for his achievement as a teacher and master craftsman. UK training manager for stonemasonry.
Martin Clarke	Chief executive of British Precast and chairman of the Modern Masonry Alliance.
Michael Driver	Architect and chief executive of the Brick Development Association.
Professor Geoff Edgell	Head of building technology at CERAM, a leading independent research and development company in the UK. Former president of the British Masonry Society.
Cliff Fudge	Technical director for H+H Celcon Ltd, manufacturer of aircrete masonry products. Former president of the British Masonry Society, he now sits on BSI masonry- and aircrete-related committees.
Dr Jacqui Glass	Lecturer in architectural engineering in the Department of Civil and Building Engineering at Loughborough University. With an extensive track record in research on innovation, process and best practice in construction, Jacqui has also worked as consultant to the cement and concrete industry and has written numerous articles and industry publications.
Dr David Hills	Sustainability manager for Ibstock Brick Ltd, the largest manufacturer of clay masonry products in the UK. Chair of the Brick Development Association's Sustainability Steering Group.
Paul Monroe	Curriculum leader of construction crafts at York College, one of the most prestigious construction craft colleges in the North of England.
Ian Pritchett	Managing director of Lime Technology Ltd, a company that sells lime-based products and is committed to creating a sustainable built environment. Ian has written numerous articles and papers on the use of lime and sustainable masonry, and lectures widely on the subject.
Professor Sue Roaf	Professor at the School of Architecture at Oxford Brookes University. Author of many well-known texts on sustainable construction.

1 DEFINING SUSTAINABLE MASONRY

How would you sum up the term 'sustainable masonry'?

Ian Abley	*Sustainability is such an elastic term, and so many charlatans have smuggled so much rubbish into the term as well, that you're better off not using the prefix 'sustainable' for anything, really. It's a gift for people who'd like to pull the wool over your eyes.*
Martin Clarke	*Oh gosh, there are lots of different ways of describing it. I mean, fundamentally, my interest in sustainable masonry is in keeping the industry sustainable. So, making sure that we preserve the skills necessary to build in masonry, making sure we've got the products and innovations.*
Michael Driver	*Long life, durability, easy maintenance, adaptability, solidity, and if you've got that, I think you've got sustainable masonry.*
Professor Geoff Edgell	*I would say that sustainable masonry is masonry construction that uses materials and processes in such a way that it does not inhibit the possibility of future generations doing the same thing.*
Cliff Fudge	*I'd see that as masonry that is designed for longevity, using materials … Basically all masonry products are indigenous materials, so in a sense they already generally have relatively good credentials in* The green guide.
Dr Jacqui Glass	*The broad definition of sustainable development in terms of economic, social and environmental objectives rather than just considering economy, that's the understanding that I have of sustainability, and so if one is applying that, therefore, to masonry and calling that sustainable masonry, then I'm happy with that as a definition, basically.*
Dr David Hills	*It'll be the fact that you've got a durable, low-maintenance product that's going to last, that's going to continue looking good, require minimum maintenance, that's going to meet the challenges of thermal performance from global warming and things like that.*
Ian Pritchett	*I think it's a balance between its performance and use. It's got to perform well, it's got to be long-living, it's got to be durable, and a balance between the amount of energy that goes into producing it in the first place compared with how long it's going to last. So, to me, masonry that's sustainable isn't just the masonry units, it's the whole building system and a mortar that allows you to take the masonry units apart and reuse them.*

I think that I would probably sum it up in the relationship between the age, the examples of the materials used in the real environment, a relationship between the age of a building and the condition in which it is found. So you might take an age of a building maybe between 100 and zero for the vertical axis and across the horizontal axis you'd put the condition of the building from zero, zero being good and 10 being rubbish or vice versa, so you could actually plot how buildings last in different construction types and the ones that last the best are more sustainable.

Professor Sue Roaf

Findings of literature review (Chapters 2–4)
Masonry that considers the use of local/reclaimed/recycled materials, longevity, thermal comfort, ease of refurbishment/extension, minimisation of embodied/operational energy and design for deconstruction.

Summary of interviewee responses
Overall, within the above definition without being as broad.

2   GOVERNMENT ATTITUDES

How do you view the government's promotion of MMC in its quest for the £60 000 house, in terms of creating sustainable buildings?

Ian Abley

The success of the MMC agenda, or its lack of success, was measured between September and December 2006, when the Code for Sustainable Homes came out and virtually everybody that was alive to what was going on stopped talking about MMC and started talking about the Code for Sustainable Homes ... the phrase 'MMC' came in with Prescott, it died as soon as Prescott died politically.

Martin Clarke

MMC was not developed with sustainability in mind at all. And neither was the £60 000 house. The £60 000 house was conceived as a cheap, I say cheap ... a political move to try and drive down the cost of a new house, make it affordable, which hasn't actually worked.

Michael Driver

I think the £60 000 house is nothing more than a good headline for the government.

Professor Geoff Edgell

We have had many, many sorts of these systems through our laboratories for testing purposes ... I think that the difficulty with many of them is that the detailing has not been sorted out properly, so the connections between panels, how you deal with openings and window-reveals and this sort of thing, has not been dealt with adequately, and I begin to wonder whether or not some of the longer-term durability is actually properly dealt with by some of the accelerated testing regimes that we use.

Cliff Fudge

I think it's totally flawed ... I think MMC is a changing fad and I actually think the term will almost disappear.

Dr Jacqui Glass

It's a prescriptive list which has no basis in science. It has no sound basis to it, so MMC, I think, is now shorthand for things which are supposedly innovative in one way or another, OK, and are not seen as old or fuddy-duddy or traditional in some respect. So I do question the whole basis for MMC.

Dr David Hills

They take the view, 'Well, OK, if we build it, in 30 or 40 years' time Building Regulations will have changed, the demands on the environment, the profile of the people living in the area, their occupations will all change, so it's going to be easier to knock down the building and start again.' But then you look at all the houses built in Victorian times that are still perfectly habitable and people are still living in quite happily, and you think, 'What's basically wrong with that?'

I think it's completely misguided. I think they've lost track of the focus. I understand where it's coming from and actually the definitions in the documents of MMC sound quite sensible, but it's all translated to off-site manufacturer, lightweight prefabricated units, and I think if we're not careful we're going to end up with a load of rubbish that doesn't last very long. — Ian Pritchett

It seems to have been based on a sort of whim and there is no demonstrated case study where it's shown to be an acceptable solution over time because we really haven't had the time in which it's been up, so you can't say it looks brilliant after 30 years, because we haven't had it up that long, so the idea of pushing everybody into an unproved technology seems to be slightly short-sighted, and the reasons why it's promoted, such as being quick to build, often don't turn out to be true in studies that I've seen. So it seems to have been promoted without much substantiation. — Professor Sue Roaf

Findings of literature review (Chapters 2–4)
Concerns relating to overheating during summer months, fire resistance and durability.

Summary of interviewee responses
Widespread scepticism of MMC, particularly with regard to durability issues.

3 MODERN METHODS OF CONSTRUCTION

Do you think that by offering its own MMC, the masonry sector is selling itself in the best fashion (ie do rapid masonry construction methods result in durable, flexible buildings)?

Martin Clarke	*I don't think it's selling itself. You know, I don't think it's a problem using thin-joint glue instead of mortar, providing that it's properly tested and properly constructed and the skills involved are different skills ... I think thin-joint aircrete is going to last hundreds of years, as other forms of masonry will.*
Michael Driver	*No, the key to good development is to make spaces and places that enhance people's lives, and unless you do that, you're wasting your time.*
Professor Geoff Edgell	*I think in general the industry should say, 'I'm sorry, we can't really go down the road of speeding up construction that much,' but I think the industry ought to focus on how it improves the production process and the construction process as well.*
Cliff Fudge	*Probably not, actually, and I think you've also got to think, 'Does the market want rapid build?'*
Dr Jacqui Glass	*The masonry industry in its broadest sense has come under fire from media and from various people ... it's old-fashioned, it's slow, it's messy, it's labour-intensive, we have a skills shortage, apparently. All of those criticisms are running towards it, so if the industry can respond to any or all of those criticisms by putting up a different type of its product and saying, 'This is a more modern way of doing things,' then of course it will do so, and it makes perfect sense to do that.*
Dr David Hills	*As far as I can see, everybody that I talk to, and I don't talk to a huge number of people, but I've talked to a lot, they all basically say, 'Brick is the material we want on the outside of the building' ... I think we've got to look at thin-joint masonry, we need to look at other things like that to possibly move away from the traditional 10 mm water joint and things like that, to get some improvements in performance.*
Ian Pritchett	*I think only time will tell. I think it's right for the masonry industry to try and respond with MMC and to look at that, but, to me, masonry has been around for thousands of years and has worked for thousands of years. Maybe trying to change it to something, whatever you call it, MMC or not, isn't necessarily the right way forward.*
Professor Sue Roaf	*I think that we do need to build fairly quickly, but I do not think that the argument that you're building it quickly is a reason for putting up something that is shoddy and won't last.*

Findings of literature review (Chapters 2–4)

Concerns that trade literature tends to concentrate on rapid build/increased production and low cost rather than durability and aesthetic appeal.

Summary of interviewee responses

Generally, the masonry sector is justified in looking to improve materials and processes, but should not necessarily be concentrating its efforts on speed of build. The comment made by Martin Clarke in relation to the durability of aircrete products was mirrored by Cliff Fudge, who during his interview with the author stated that the first aircrete products were used in 1924 and that the buildings on which they were used are still standing and have required little maintenance.

4 WHOLE LIFE COSTS

When considering that MMC has proved to be more expensive than masonry construction, wouldn't it make more sense for the masonry sector to promote the consideration of WLC and the development of high-quality products, which take longer to build (ie more traditional products)?

Ian Abley

You can look at a stock of 26 million homes in Britain and if they all lasted 100 years, you've got to build 260 000 of those a year. If they all lasted 50 years, 10 years shorter than the BRE figure for MMC ... they've got to build 520 000. If they last 25 years, you've got to build 1 000 040 ... But then on the old Loss Prevention Standard 2020, which is the definition of MMC structures, the structure only has to last 20 years, but the cladding can last less. If your aesthetic then has to be replaced because the steel sheet it's sitting on, clipped onto, needs replacing, you are going to have a million of those façades rebuilt every year. So the maths just goes completely wrong.

Martin Clarke

To government the word 'traditional' has become associated with unsafe, with dirty, with inefficient, with overproduction of site waste, the old way of doing things. The word 'traditional' has become, temporarily I think, a negative, and that's another way, another reason why we adopted the word 'modern', it's just the flip-side of the same coin ... Most small builders actually like the word 'traditional', they quite often use it in their marketing information. Not so the majors, the majors are much more volume-orientated and they don't particularly want the consumer to know how a house is built.

Michael Driver

Yes, but you see, the snag is that at the moment the government is obsessed with this idea, and therefore at the moment you are being dragooned into doing that to such an extent that they won't sell you the land unless you're going to do a certain amount of MMC on it. So it's ... yes, I mean, in the long term, quality, durability is what's important.

Professor Geoff Edgell

I think a little bit of evolution here and there ought to be the way that we move things forward.

Dr Jacqui Glass

Well, I think to some extent they are making that case ... the masonry and the concrete industry have been quite vociferous in calling for a longer lifetime assessment period ... what they haven't got is a really robust case and really robust data. I don't think anybody has really got to grips with that, in terms of actually a proper study which shows that.

It's got something that works compared with something that maybe doesn't work, but the one caveat to that is actually, masonry works in structural terms, it works in longevity terms, it works to an extent in comfort terms, but it isn't always totally comfortable or energy efficient, and so there's got to be a change somewhere because we're now faced with a need to start to build houses that don't need energy or need a lot less energy. Solid masonry or cavity masonry units in their existing forms don't meet all of those needs, so something's got to change, so I think it would be wrong for the industry to say, 'We've got everything right and we don't need to change.' I think those things need to be explored.

Ian Pritchett

The challenge now is to provide extremely robust and resilient structures that can withstand the exigencies of a rapidly changing climate – number one.

Professor Sue Roaf

Findings of literature review (Chapters 2–4)
The masonry sector needs to evolve, but should work on the strengths that have been the basis of its success for thousands of years and not bow to government pressure to do otherwise.

Summary of interviewee responses
As per literature review. Interesting points raised relating to the lack of WLC information and the fact that house builders do not provide enough information on their products.

5   ENERGY IN MANUFACTURE

In what ways are manufacturers of masonry products looking to further reduce energy during production and the drain on fresh resources? In terms of such improvements, how much further can they go?
(Note: Each interviewee asked this question has an industry-leading specialisation in a particular product, which is shown in parentheses next to their name.)

Cliff Fudge (concrete blocks)	*We've decided that, in terms of water, we would try and limit to an absolute minimum the amount of water taken from mains. And we either use canal water close by … but the biggest thing we've done is, it's quite a big site and it's paved, all the rain water from that site goes into a lagoon, and they clean it and use it in the process, and then return the water back into the lagoon wherever possible. We try to minimise the high embodied energy materials, you know, like cement … we've got CHP, combined heat and power, on our latest plant as well … The aircrete industry predominantly uses fly ash, waste by-product from power plants, but we're also looking at other forms of waste material that don't give you a product in the process of course or in the end product. We look at any cooked waste, as we call it, grinding it down and either making an anchor-type block from it or grinding it back into the process so that we can get it completely recycled … there's a lot that can be done.*
Dr Jacqui Glass (cement)	*They made a massive improvement in their energy efficiency in kilns by changing the wet process to a dry process, so a lot of new kilns and improved kilns in the UK have effectively had massive engineering changes to them to do that. They've also looked at fuel. Rather than using all coal … things like waste products like tyres, solvents, papers, all those sorts of things … and obviously if you can also look at your emissions from the chimneys, if you can do heat recovery from your chimneys … Castle Cement's new kiln at the Padeswood Works was designed … to take all these different fuels, but its planning permission, as it stands, doesn't allow it to burn all of those … It's becoming less economical to make cement in the UK, and they're starting to bring it in from overseas … They know they can go further.*

We [Ibstock Brick Ltd] spent £55 million. There were three big developments amongst that … We monitor the energy that we use very, very, very carefully … We will hit a 10% reduction in the energy saving by 2010, based on the 1990 benchmark, so that's a 10% saving … kiln technology is a big part of it … landfill gas, electricity generation … we have an active programme of development, we are looking at increasing perforation to reduce the amount of clay, so we can cut down the energy consumption to a certain amount of burning technology. With mineral replacement as well, what figure could we achieve? That's a tough one. I know we're going to exceed 10%, but how much above 10% I don't know. We might get as far as about 15% in terms of CO_2 emissions. And then you've really got to ask yourself some questions about … Do you then go on to think about carbon trading? Do we then either trade or do we offset or whatever?

Dr David Hills (clay bricks)

Findings of literature review (Chapters 2–4)
Manufacturers of masonry products have made great strides in recent decades to reduce their energy consumption and the use of fresh resources.

Summary of interviewee responses
It seems clear that, although more can be done, barring a massive technological breakthrough manufacturers are nearing the uppermost threshold in terms of energy efficiency. Interesting comments in relation to planners holding back the use of alternative kiln fuels and water efficiency, which was not considered in great detail during the literature review.

6   RECYCLABILITY/RECLAIMABILITY

It is often stated that masonry construction is fully recyclable/reclaimable at the end of its useful life. Is enough being done to ensure future reclaimability and, if not, what can be done to aid the use of reclaimed materials?

Ian Abley	*Reasonable foresight, which is the only test that the law demands, it doesn't require architects to be clairvoyant, it requires architects to have reasonable foresight, anticipating that maybe your client would want to knock down that building and recover the brickwork, that's a legitimate thing, but trying to fossilise society through its buildings died out with the Egyptians.*
Martin Clarke	*Yeah. I think that's a good road for us to move down. I think the old lime-mortared brick walls are the source of most reclaimed brick, because you can knock the mortar off and get quite a nice clean brick out of it.*
Michael Driver	*The problem with reclamation at the moment, of course, is that they are reclaiming stuff which we've got no control over at all. Once they start reclaiming buildings which have got an identified breed, then there is no problem … but at the moment it isn't high on the priority list.*
Professor Geoff Edgell	*Well, I'm sure it hasn't … Personally, I don't think it would be too difficult to look at different ways of texturing the surface. It may just be, if you imagine, if you look at wire cut bricks, where you're extruding the column and all you do is cut it into a number of units with a series of wires, though it may be that by varying the shape and so on, the cross-section of the wires that you use, it may be that you can create a bedding surface which actually initially sticks extremely well to the mortar, but under the action of a tool for recycling the unit, actually comes off. It seems to me that it shouldn't be beyond our wits to come up with something that could do that.*
Cliff Fudge	*I think certainly more can be done … We're starting to think about huge units, big, big wall elements that are put together dry-laid, that could be demountable and reused, but in masonry in its current form you need the bond in between the brick and the mortar for the masonry to work as a whole.*
Dr Jacqui Glass	*It's the mortar that's the problem … I suppose you're saying that one can only get a certain credit in the Code for Sustainable Homes if you specify reclaimed bricks, or if you specify a brick that can be reclaimed more easily then demand would rise and therefore the manufacturers would have to respond. So I think to some extent it takes that sort of driver to make that change.*

(When pressed on whether Ibstock Brick Ltd could produce bricks with a surface texture that would aid reclamation.) Yeah. If somebody was to say to us at the outset, 'We want to build a building that is a sustainable building and we want to recycle it at the end' … But at the moment that's not on the radar screen … At the moment, unless the designer makes that decision at the outset, it restricts your options.	Dr David Hills
No … Masonry is only fully recyclable as long as you don't use mortars that are too strong. At the moment, the engineering side of the construction industry always adds a few extra safety factors on, which means you end up with mortars that are a little bit too strong … I think that the steps we now have for factory manufacturing of mortars mean that you can have confidence that what you're specifying is what you're getting, which means that you can use lower partial factors of safety in the specification of the mortar, and should mean that you can recycle more of it.	Ian Pritchett
The cost of materials is going through the roof now, isn't it? You're going to get to a stage now where all bricks are recyclable, don't you think?	Professor Sue Roaf

Findings of literature review (Chapters 2–4)
Not enough being done. Mortars too strong. As-built information needed.

Summary of interviewee responses
As per literature review. Very interesting admissions that masonry units could be manufactured to aid reclamation. Lack of driver for reclamation.

7 LIFESPAN

Brand (*How buildings learn*, 1994) suggests that the structure of a building should be designed to last up to 300 years. What should the target be for today's masonry buildings?

Ian Abley	*I'd agree with the idea of long life, loose fit, low maintenance, but to pin a number on the long life, and to equate the long life with the potential, as if it was embodied in the material, I think that's a mistake, and the idea of adaptability or flexibility or disposability or extendibility, all those abilities, I think they're all good things … but just because glass lasts 2000 years doesn't mean the building should.*
Martin Clarke	*There is no process of deterioration if a house is built properly, it should be there for a few hundred years.*
Michael Driver	*I think if they are making spaces and places that enhance people's lives there's no reason why they shouldn't go longer than that. And in fact, statistically, they've got to, haven't they, to cope with our replacement rates? I believe that a house that's built now, statistically has got to last for about 750 years to even make its self worth.*
Professor Geoff Edgell	*I think the key to this is they should be designed to last up to 300 years, and as a statement I don't think that's unreasonable.*
Cliff Fudge	*I think if you told a house owner, 'After one term of mortgage, your house is worthless' – as they do in Japan – or even after two terms of mortgage, then the philosophy in this country is not, 'Oh, you know, it's a diminishing return,' they actually think that this is an investment and it will actually last a long time, so hundreds of years, hundreds of years, I think, is a very simple way of putting it.*
Dr Jacqui Glass	*Between 60 and a hundred-and-something years. If you're not playing in that ball game, it starts to become madness, really, as something which is triflingly inadequate or substantially almost over-specified and specified for all the sort of wear of an Oxford college over 400 years. So I think it's got to be in that range somewhere, yeah.*
Dr David Hills	*It's the designer that has really got to make a lot of these decisions, and at the moment I think designers are far more focused on working capital, construction time and profit margins and they're not looking at the big picture. But things will change.*

I think we should be designing for 300-year life cycles, we should be assessing when it comes to full life cost and full life-cycle assessment of a building, we should be assessing it on the basis of its true life or its estimated true life, rather than an arbitrary 60-year figure, because if something's going to last 60 years as against something else that's going to last 200 years, you assess them both on a 60-year life then you're giving a false impression, aren't you?

Ian Pritchett

Yeah, I've seen all of that, yeah. Brand. Yeah. Three hundred years, him and me, just like that. (Crosses fingers.)

Professor Sue Roaf

Findings of literature review (Chapters 2–4)
Longevity and masonry go hand in hand. History has proved that 300 years is not an unreasonable lifespan, but there are concerns that cavity wall construction is not as durable as solid wall construction, which was widely used until the mid-1940s.

Summary of interviewee responses
Generally, that masonry should be designed to last as long as possible. A few comments that a building's lifespan should be based upon the client's needs/use of the building, which would again seem to raise the issue of ease of deconstruction if this is lower than the life expectancy of the masonry units.

8 SOLID WALL CONSTRUCTION

Due to increasing concerns over buildability issues resulting from recent changes to Part L of the Building Regulations, some are advocating the reconsideration of solid wall construction. Do you think that the masonry sector has the capability to produce waterproof solid wall construction with aesthetic appeal?

Ian Abley

Yes. Solid wall construction could start coming back in all sorts of different forms, really, where you get sandwiches, much like concrete sandwich panels have now become composites, but they might have insulation sandwiched in between two leaves of concrete, rather than just being a solid lump of concrete.

Martin Clarke

Yeah, I think so, particularly with aircrete where the buildability is more easy than cavity wall construction. It can be done, is done, yeah … there's fight for the outside wall between brick, between clay brick, between render, between reconstructed stone, cast stone, between natural stone.

Michael Driver

For the best insulation you insulate externally. There's no argument … the cavity, which is basically … you've got some sort of form of building here, you've got a cavity, then you've got your external skin. I mean, I think that's a very logical way of getting round that problem.

Professor Geoff Edgell

Well, I think that the durability of the sort of construction you're talking about is probably less than that of a solid wall … whereas a modern cavity wall, where you've built in some mineral wall slabs, for example, into a cavity, I think it's fairly inevitable that over the life of the building in some areas it's going to get damp and dry out and damp and dry out, and I think it's very likely that it will deteriorate and, as you say, if you're going to do anything about that, it probably is a fairly major piece of work. You're going to have to start taking down the inner leaf and pulling things out or do it from the outside. So I think there are significant built-in problems there.

Cliff Fudge

I would say a big 'Yes', actually … I suppose, just thinking in terms of robustness, structural robustness, solid wall has got to be the best option, because it's easy to construct and easy to design … You may get settlement on insulation in the cavity with other forms, it could get wet and its values could deteriorate.

The research committee that I chair has. Ibstock, I can't tell you what Ibstock's doing, unfortunately, because that's confidential again, but we have looked at it, we have got ideas, we are contemplating a number of issues on wall design per se, so we are looking at all different types of structures, and, yes … With the modern types of insulation, high-performance insulation and with the ability to have a waterproof membrane, we're pretty confident this solid wall could do it. I can't be more specific than that, Mark.	Dr David Hills
I think there are ways of doing it, but they're not what you would call conventional masonry. There are things that we're working on that are based on lime and hemp bio composite materials which we view as masonry, solid wall monolithic masonry, or it can be in block form, and we're looking at systems there where they can be faced with bricks or used with conventional brick … I think it will work, it can work, but there's going to be a little bit of R&D.	Ian Pritchett

Findings of literature review (Chapters 2–4)
In view of rapidly changing requirements for levels of thermal performance and the fact that insulation is likely to deteriorate in a cavity, which poses replacement problems, the benefits of solid wall masonry should be reinvestigated.

Summary of interviewee responses
General agreement that solid wall construction should be reinvestigated. Manufacturers of different masonry materials are in competition in this sense when cooperation and the combination of different materials could prove to be the best way forward. General consensus that the location of wall insulation is a massive problem that the masonry sector needs to consider seriously. Ian Abley, as part of his doctorate work for MMA, is examining the possibilities for the use of vacuum-insulated panels.

9 PREFABRICATION VERSUS BETTER SITE CONDITIONS

Is prefabrication a possible improvement or should the masonry sector be promoting the creation of better site conditions that are conducive to good masonry?

Ian Abley

Site conditions definitely. Paying bricklayers more money, that's the first thing. Craftsmanship is the thing that's grotesquely undervalued, and the idea that you get … a bloke has got to pay the mortgage at the end of the week and lay a thousand bricks and still make it look fantastic, is … in any weather … it just floors me.

Martin Clarke

Well, both, yeah, I would say prefabrication of masonry, it's if you want to tender for an MMC scheme without thin joint, then you've got to find a way of getting that masonry on in a prefabricated pattern. If you can make more money out of doing that, then so be it, I suppose … Anything that's prefabricated, I mean, one of the problems with MMC systems is they get the stuff to site and it doesn't fit.

Michael Driver

Well, we are trying to do that – better site conditions. I'm also the chief executive for the Association of Brickwork Contractors, the 24 biggest brickwork subcontractors, and the standards to which those people work are very high … There is a problem insofar that in the mass housing market, people hire and fire brickies and they use them as cannon fodder, and until people accept that there are trained, skilled tradesmen who ought to be having proper conditions to work in and everything else, then you've had it.

Professor Geoff Edgell

Yeah, I think that there could be a lot better site practice, which would lead to a lot less snagging difficulties at the start, would lead to a hell of a lot less waste … I think there is, not necessarily in housing, but I think there is a role for pre-assembly, so I'm thinking of a situation, but it's largely commercial, where you might have repetitive detail, say some columns, which could all be built somewhere off site, maybe then slightly pre-stressed … I think there are situations where that's a right way forward.

Cliff Fudge

I really think – we discussed this yesterday, funnily enough in our meeting – I think you've got to be looking for improved on-site construction. Off-site construction has its benefits, but it has a lot of downfalls and the number of sites we've been involved in recently where you start the job and there's a last-minute change to design … You can deal with a lot of irregularities in terms of site layouts and so on with masonry, which you can't do with a fixed-panel system.

They [prefabricated panels] all give you a problem somewhere on the structure, that if you want to do the side of a supermarket, no problem, but you go round the corner you change elevation, you change height ... anything like that, any slight change in detail, in design, you end up somewhere with a problem that you've got to find an almost unique solution to ... many of our customers are still not convinced, they still go the traditional way.

Dr David Hills

Perhaps there should be more of a campaign to tell kids, 'If you want to be a bricklayer, don't work for a cowboy' ... You should be demanding a warm place to have your lunch, you should be demanding a place to put wet boots, you should be demanding a warm room to hang your coat up in.

Paul Monroe

I think there's room for all of these things. Obviously we need to improve site conditions. I'm not a great fan of prefabricated masonry, but it has got its place. There are certain elements; one of the obvious things is arches work quite well being prefabricated, don't they?

Ian Pritchett

Findings of literature review (Chapters 2–4)
Excellent results can be achieved by ensuring better site conditions.

Summary of interviewee responses
As per literature review. A number of concerns were raised over tolerances on site, but it was widely accepted that prefabrication has its place (ie the production of specialised elements). Strong comments about the poor way in which tradespeople are perceived and treated.

10 SKILLS SHORTAGES

Advocates of MMC suggest that changes to training regimes have resulted in a shortage of skilled tradespeople. What are your views on such comments?

Ian Abley

Often the worst payers will argue there's a skill shortage, because they're just not willing to pay for them, so there's always been that 'I can't get the staff' kind of complaint … It's a bit of a myth, really, but you have got to commit to skills and you've also got to commit to the blokes that are on … you want to retain, so the better companies will be the ones that do train up their guys … Round our way they're building these academies, so my son, who's seven now, by the time he's 11 there should be an academy up the road, and it's funded by KPMG, and it's an academy that's supposed to be devoted to business finance and maths as well, working in a bank, filling out mortgage forms, all that, but where are the academies for masonry? That's what's needed.

Michael Burdett

There's a lot of financial constraints put onto the learners. They've got to achieve so many other things outside what would be their normal craft, they've also got key skills and there are other bits and pieces and also they're squeezing the training time all the time. You're looking at, for a level 2, they've got 23 weeks in college to get from being a new entrant with no experience whatsoever, to going out … that deems them competent to be a skilled craftsperson in the industry.

Kevin Calpin

If you go back to 1993 there were two million people in the construction industry … Due to the recession in the late Thatcher years it dropped to about one million people, so automatically there's a shortage of skilled tradespeople. House building, I think, is the question. When they come out, we are building it back up, but colleges only have a limited capacity. Our workshops are full to overflowing every day … and still the country needs more people because the money isn't being put into the training provision. Give us more staff, give us more space, give us more materials and they can have what they want, but at the moment they're asking us to do things that are impossible.

Martin Clarke

Don't think so. No, I think that was hogwash, actually, I think that was a scare story that was put about by the timber frame industry, and always will be … I think a bigger challenge for us is that it's not so much numbers, it's making sure that people who we attract into the industry enjoy a safer environment, are properly paid and properly looked after and not treated as casuals.

I don't believe it. Yeah, I don't believe it … I'll just take our industry figures, apply to everybody, we manufacture about three million cubic metres of aircrete per annum, we've been doing that for round about 20 years … We've had a problem with bricklayers holding us up.   Cliff Fudge

With modularisation, off-site prefabrication, thin-bed technology, more construction plan, more use of power tools, all these things, and I think, without any criticism directly of NVQs, inevitably that leads to a lower skill level. Not necessarily of every tradesperson, because there may be the odd young man or young woman who gets hold of skills and can work in traditional or refurbishment or heritage.   Paul Monroe

We should be looking far more to vocational skills. Sadly, we're all in the business of needing, as a material producer and supplier, we're looking at trying to de-skill the use of our materials because there is a de-skilled industry, but I'm not quite sure it's as bad as people make out.   Ian Pritchett

Findings of literature review (Chapters 2–4)
No clear-cut evidence of a shortage of skilled craftspeople.

Summary of interviewee responses
As per literature review, but again, strong criticism that some companies treat tradespeople poorly and are not prepared to pay reasonable rates for high skill levels. Lecturers feel that today's training regimes are too restrictive in terms of time and resources.

11 TRAINING

Do you think that today's training regimes equip trainees to cope with the production of high-quality traditional brickwork/stonework?

Michael Burdett

The very specific tasks such as English bond, Flemish bond, yes, I'm fairly confident that the level 2 learners could go and do that, but anything outside that, probably not. Our route splits two ways: you've got craft masonry and a bricklaying route. The craftmason tends to follow a general building, which looks at the roofing, plastering and all the other things you would expect from a bricklayer as well, but there isn't much demand for that because everybody likes the bricklaying route. But again, it's very specific, because if you split it down further into the units, we go back to the common core, the six units to get the NVQ, three of them are core units and then three of them are made up of a setting out unit, the actual bricklaying craft unit and then there's an optional one, which may fall into drainage or concrete, etc. But it's quite specific in what they've actually got to do, it's not the general ... For example, from a bricklayer's point of view there's nothing, no mention in the scheme whatsoever about foundations. When I served my time you started at the bottom and you worked your way right through the building process ... Now the learners are going out with only very specific knowledge.

Kevin Calpin

The skills are there. What you don't get, and we spoke about this from the very first couple of minutes, about people wanting to stretch themselves, but there are still people like that about. Mike [Burdett] and I grew up on that, because Mike is UK skills technical training manager for internationals, and I do the same for stonemasonry, so we know, even within our own colleges, there's a nucleus, we've got some kids who really want to get on and do things, and they're going to be the ones who win the ... they come to our presentation, they'll win something, they'll go into the competition, they'll be representing York College and there's that nucleus in colleges all over the country, and that's what we have to work with for an international.

One thing that you have to know if you don't know much is ... there is no NVQ in carpentry and joinery. It became an NVQ in wood occupations, which had many different forms, including shop-fitting, but now you can do an NVQ either in site carpentry or bench joinery, so there seems to be one or the other. You can even do an NVQ in painting, or you can do an NVQ in decorating, so you would have painter-decorators. You can do an NVQ in hard plastering or an NVQ in dry lining or an NVQ in fibrous plastering. So what they're doing is chopping it up all the time. Like, 'Mum, Mum, could you make me a bacon sandwich?' 'Sorry, son, I'm only trained to do egg sandwiches, or sausage sandwiches, but I'm not allowed to do bacon sandwiches.' And that's pretty much how it's gone. Now, that allows them to get the on-site evidence, which is what they want, and it allows them to do it in the time, but you wonder whether the qualification was written for the kid or for the company.

Paul Monroe

Findings of literature review (Chapters 2–4)
Data not available for literature review.

Summary of interviewee responses
Lecturers expressed concerns that rather than providing a rounded craft education, NVQs have tended to divide trades into a series of skills and that it is very much left to the individual student to decide how far they decide to progress their own skill base.

12 GOVERNMENT SUPPORT

Is the UK government doing enough to encourage sustainable construction?

Ian Abley

They're doing too much. They should stop and there should be a ban on the word 'sustainable'. It's not doing enough to promote construction, in the sense of production and output, it's doing loads of R&D, but that R&D is weakly connected to the process of actually getting stuff built.

Martin Clarke

Well, I don't know. Probably not enough, no, it's not really actively saying, '[This is] best practice, client' … You can do an awful lot more by showing the way.

Michael Driver

It's doing so much about it that it doesn't know what it's doing.

Professor Geoff Edgell

No, I don't think it is. I was in a meeting in London on Wednesday, an organisation that's now a year old called Materials UK, and one of the guys there, who's now working for [the Department for] Communities and Local Government, came forward and said, 'I'm working on this sustainable construction strategy,' and he listed out things that were important in developing a sustainable construction strategy, but under each heading there were probably seven or eight other initiatives already running … You start to think, 'God, if we do all these, by the time we get round to agreeing what this strategy is, it'll take five years to review all the initiatives that have already taken place' … What we don't have now is a centre of excellence for knowledge in building construction in the UK, a government-funded body.

Cliff Fudge

But I think the government's very clear where it's going and you can't turn over any page in a trade magazine without finding something about construction without sustainable construction and sustainability.

Dr Jacqui Glass

No. No, it's not. It's … I was thinking about the questions last night and I was just thinking, 'How do I actually sum up all the things I think about this?' The government is not doing enough to promote sustainability, sustainable development principles on the whole way of British life … You start to pick up things like the Code for Sustainable Homes, and you think, 'Actually, this could make a lot of difference,' but of course, as you probably well know, it was supposed to be the Code for Sustainable Buildings, not just homes, not just on government land, and it could have … it has no teeth.

No is the quick answer, because actually the Code for Sustainable Homes is distorted in perspective ... The Code for Sustainable Homes has been launched, but it's nothing to do with sustainability and it should have been called the Code for Low-energy Housing, something like that ... If you're going to use brick, then use brick for the virtues that it's got and the longevity and the low maintenance and things like that, and if you work out that footprint per square metre per year of life, you're going to find that – and I've done the calculations – you're going to find that clay brick is far more efficient and far cheaper with a smaller carbon footprint than many other materials, purely because of maintenance and replacement.

Dr David Hills

I think they're doing a huge amount. Whether it's enough, I have to say I'm incredibly impressed at the amount of stuff that's going on, even though I don't agree with everything or think that it's all necessarily right yet.

Ian Pritchett

Well, I don't know how they're particularly doing it, do you? I think we ought to be having discussions about sustainable developments around a table at which the interests of local communities are given the same weight as the interests of industry.

Professor Sue Roaf

Findings of literature review (Chapters 2–4)
The widespread promotion of MMC conflicts considerably with many aspects of what commentators normally regard as 'sustainable construction' (ie longevity and design for climate change).

Summary of interviewee responses
Some praise for the amount of work being done, but a general consensus that there is no clear driver or way forward and that a lot of guidance/information is conflicting.

13 PLANNING LEGISLATION

Does planning legislation do enough to ensure the optimisation of durability, enduring design, the use of reclaimed materials, and that developments are laid out to optimise their ability to create an acceptable internal environment with minimum energy use?

Ian Abley

No, because it can't. What is planning? That's the question … It isn't obvious why building control and planning in Britain evolved to separate regimes at all. Because if it was planning for production, an 'I'll get out of bed in the morning and put this pair of trousers with that shirt' kind of planning, a deliberate effort to end up at the end of the day with something made or achieved, then planning for production would have related building control to the permission to carry on. But planning came after building control … and came up with byelaws and just the improvement agenda of the Victorians, really, sewers and lighting and just wanting to make things better, stop disease. Planning landed on Britain as a denial of development rights and an attempt to stop people building. Albeit for high-minded and well-intentioned reasons, no doubt, they wanted to rebuild the country and build council houses, but it was a restriction and a negative thing, and they tried to make it more positive since, pretty well constantly … the introduction of consultation in the '70s, the third party … they tried to fix planning as a denial of rights, every decade since in that last 60 years. But I think it's just got to the point where, if they really want to push the technology and building production to higher levels of building performance, they can't now have a separate system. It just doesn't work.

Michael Driver

This is a real problem, because at the moment I believe that what's going to happen is that when the Code for Sustainable Homes really starts to bite, you will find more and more firms of architects and designers getting involved in the development of this, and when that happens you're likely to see houses that change. I don't see that housing has got to have any dramatic changes, we're still the same height, we still need to lie in a bed, and all the other … we still need to wash, etc., but I do think that people are going to come down to sort of … You can look at Swedish developments, way and beyond the actual house itself, I mean, the way that they have the central heating production, central waste disposal, all those sorts of things, and that's really the level at which I think the housing world is going to change.

Dr Jacqui Glass

Broadly, no. Probably in some specific instances, yes, it does. I think it very much depends on the local authority … Unless you've got someone like the GLA with Ken saying … Ken Livingstone saying, 'Right, these developments are all going to have 10% renewables in them,' or whatever, unless they've got that edict, they won't go near that sort of thing.

I think some planning authorities are now moving in that way, but it's not only the planning authorities, it's the people who sit on the committees that pass it.

Professor Sue Roaf

Findings of literature review (Chapters 2–4)
Only a handful of local authority planning bodies currently actively encourage sustainable development. Recent moves towards zero-carbon buildings have set goals that might best be met by combining the skills of planning and building control officers more frequently than is currently the case.

Summary of interviewee responses
As per literature review. Suggestions that a new approach to 'development control' may evolve.

5.3 SUMMARY
This chapter conveys only a fraction of the immensely valuable information gathered during the semi-structured interviews. Although a few of the interviewees' opinions differed somewhat markedly from those expressed by the majority in a number of cases, overall the interview data served to confirm the findings of the extensive literature review, and in some instances added considerably to those findings.

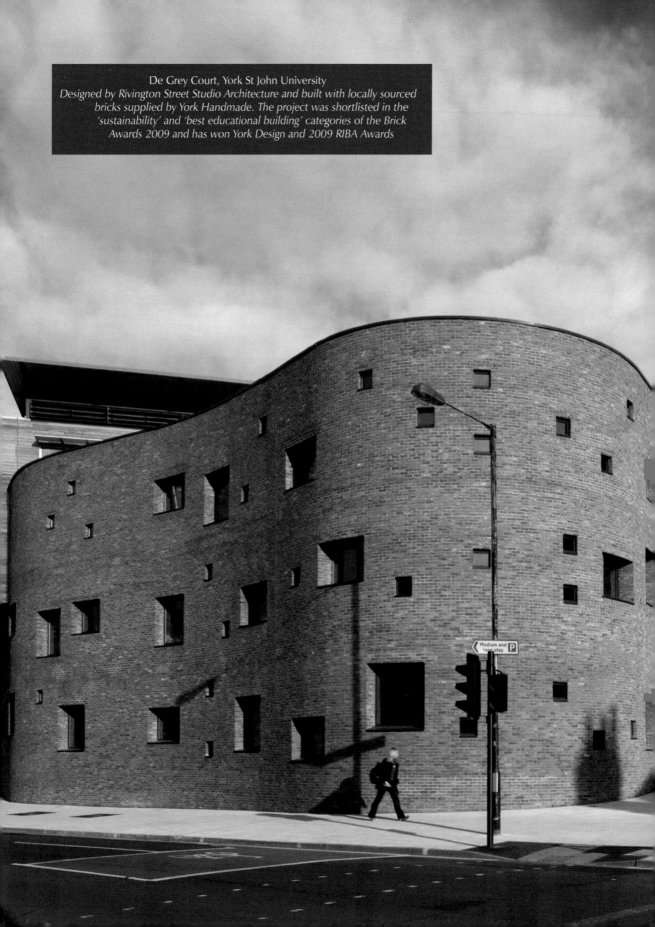

De Grey Court, York St John University
Designed by Rivington Street Studio Architecture and built with locally sourced bricks supplied by York Handmade. The project was shortlisted in the 'sustainability' and 'best educational building' categories of the Brick Awards 2009 and has won York Design and 2009 RIBA Awards

6 SUSTAINABLE MASONRY: A DIAGNOSIS

Sustainability of the built environment requires multidisciplinary, international and interdisciplinary work over many decades using both technical and humanistic approaches. Sustainable development needs a living language that is readily understood by all people. It also calls for an ethical stance and, very often, the confidence to depart from the norm. This places 'design' – by planners, developers, architects, engineers, constructors, users and manufacturers – at the centre of a process that is understandable and holistic and focuses human ingenuity.

Institution of Structural Engineers (1999, p. 8)

6.1 INTRODUCTION

Chapter 5 analysed the data from the semi-structured interviews and recorded the most relevant thoughts and feelings of the 12 interviewees, drawn from questions intended to bring out each individual's expertise. This chapter now brings this analysis together with the review of literature outlined in Chapters 2–4 in an attempt to focus the findings of both aspects of the overall research.

As with the previous chapter, the sections that follow cover the main points to emerge from the analysis of all available information and again follow the order of Chapters 2–4.

6.2 QUALITY OR QUANTITY?

Driven by the need to construct the vast numbers of new dwellings desperately required in some parts of the UK (primarily in the South-East of England), the UK government's widespread promotion of MMC has, so far, proved to be a flawed and extremely short-sighted response to the problem in hand when set against the future long-term requirements for completed buildings in this country. Although MMC has its place, particularly for commercial purposes where the business activity of an organisation results in constant substantial changes to accommodation requirements (ie retail outlets), quick completion and lightweight solutions will not, on the whole, result in the most durable, adaptable, aesthetically pleasing, thermally comfortable or, overall, sustainable solution. In turn, this has forced the masonry sector to come forward with ways to counter what it considers to be a major threat to its future, with the danger that some masonry innovations might be considered neither good long-term solutions nor sustainable.

Traditional masonry has proved its worth over many thousands of years and has many fine attributes. No one would advocate that any industry stands still in time, but it is somewhat alarming that instead of building sensibly on its past successes, the masonry sector should be pushed down the wrong road by those who should be promoting the achievement of overall excellence – including considerations such as the use of local materials, aesthetic appeal and vernacular, durability, thermal comfort, future adaptability and eventual reclamation – and not something that matches a weak political agenda.

6.3 NEW MASONRY COMPONENTS

It seems clear that manufacturers of masonry materials are doing much to reduce the energy and virgin materials used in their production, and have spent many millions of pounds to do so. This has proved to be extremely prudent in terms of reducing fuel bills and raising corporate profiles.

Manufacturers of clay masonry units and cement have invested heavily in new kiln technology. Both have examined ways in which waste and the drain on fresh resources can be reduced through the use of alternative or waste fuels and the inclusion of waste/recyclate in the makeup of products. An interesting point to come out of the interview with Dr Jacqui Glass of Loughborough University was that planning bodies are reluctant to give permission for cement kilns using alternative/waste fuels. This would appear to be counterproductive from an environmental point of view as the high temperatures involved in cement production guarantee total incineration and the amount of high-level (in terms of toxicity) waste being sent to landfill could be drastically reduced.

As little as 40% of the energy required to manufacture clay products goes into the creation of calcium silicate bricks, and it is possible to include waste materials such as blast furnace slag, china clay sand and spent oil shale as a percentage of the matrix used during their manufacture, making calcium silicate an attractive option in terms of impact upon the environment.

Manufacturers of concrete masonry units (particularly aircrete products) are in the best position to include waste/recyclate in the makeup of their products and have done so fairly extensively (up to 80% in fact). They too have looked widely at the reduction of energy use, with Cliff Fudge of H+H Celcon Ltd also discussing some interesting ways in which the use of fresh water can be reduced.

It is unclear how much further some manufacturers can go in terms of reducing the impact their activities have on the environment. From the interviewees, only Dr David Hills (Ibstock Brick Ltd) would speculate, suggesting that his company could perhaps look at reducing its CO_2 emissions by a further 5%. It is obvious that further detailed research is required in this area.

Little information is available on the long-term performance of insulation materials, which is hardly surprising considering that the industry has only had a real need for these products since the 1970s. However, their rise to prominence since the mid-1990s has been dramatic and there is absolutely no doubt that this will continue to be the case. At the moment, commentators on sustainable construction advocate the use of mineral fibre products due to the fact that they require less energy in production than cellular plastics and are recyclable at the end of their lifespan. The second of these advantages does raise a serious question: If the lifespan of mineral fibre is around 60 years, and it is placed in a cavity in slab form, how does one extract it from the cavity after 60 years to recycle it? Demands for levels of airtightness will increase dramatically in the coming decades, which also raises serious questions against the future use of cavity wall construction.

The fact that masonry can last so long is often forgotten (especially by advocates of MMC) when considering the levels of embodied energy associated with its materials, which, if taken over a few hundred years, would be minimal. Embodied energy is a new and complicated science that requires far more research before accurate comparisons can be drawn across the wide range of construction components available today.

6.4 DESIGN CONSIDERATIONS
6.4.1 Design for deconstruction

There is no doubt that the possible reuse of the vast majority of the perfectly good masonry units being sent to landfill today is being hampered by the lack of accurate historical data on the properties of such materials, coupled with high costs in comparison with those for new materials.

It is clear that to stop this ongoing vicious circle of 'produce – build – demolish – landfill', the industry must start to compile and archive extensive as-built information for future generations. However, in the case of the possible reuse of masonry units, this would be pointless unless the masonry were to be built in the first instance with reclamation in mind. Perhaps the most interesting point to come out of the semi-structured interviews was that two prominent interviewees with expertise relevant to the question posed (pp. 100–101) stated that it would be possible to manufacture masonry units with bedding surfaces that would aid eventual reclamation. That said, hydraulic lime- and cement-based masonry binders that would facilitate the reclamation of masonry units are already readily available; they simply are not used often enough. The masonry sector must be well aware of this, but has done little to advertise the fact.

In terms of specifying mortars that are both functional and exhibit an ideal bond strength to allow the eventual reclamation of masonry units, pre-mixed masonry cements containing calcium limes and air-entraining agents would seem a natural way forward. This concept is now being adopted by UK cement manufacturers and is similar in nature to the durable general purpose mix prescribed by BRE as far back as 1993.

Nobody can accurately predict what the future needs of society will be, but it would seem sensible to suggest that designers should always at least attempt to create 'scenario buffered buildings', looking as far ahead as their imagination can take them. If it is simple to create masonry that is easily reclaimable and this has no effect on cost or the overall build process, why not at least give future generations the option of reclamation? Buildings guzzled (and were designed to guzzle) energy as recently as the 1970s. Now we are faced with an energy crisis and a pressing desire to reduce CO_2 emissions.

6.4.2 Design for longevity and the consideration of whole life costs

The initial reason for the development of cavity wall construction (ie the need to reduce water ingress into buildings) has been dramatically overtaken by using the cavity to accommodate insulation. This being the case, the construction industry does not seem to have at any time paused to consider the implications of complicating the cavity to meet the rising requirements for energy efficiency or, in fact, to consider other options.

Relatively few years have passed since cavity construction began to widely replace solid wall construction, which had previously been used for thousands of years. Already, the number of building defects associated with cavity walls suggests that it will not serve society as well as more traditional methods of construction. The fact that rigid sheet insulation materials with low or unknown durability thresholds, which cannot be removed/replaced without totally dismantling a building, continue to be placed between two leaves of increasingly separated (now up to 300 mm) masonry should be an issue of grave concern. One could compare this to car manufacturers (often portrayed as the manufacturing benchmark) permanently welding in components that are known to require replacement before the body encompassing them deteriorates. A simpler, more standardised approach would not only aid longevity, but would drastically reduce waste levels. This fact, and Brand's recommendation that the structure of a building should be designed to last for up to 300 years, were widely accepted by industry experts.

6.4.3 Masonry quality and craftsmanship: do they still exist?

Despite claims made to the contrary by advocates of MMC, the resounding answer to this question is yes. Interviewees made the point that although not accepted as a wholesale masonry solution, masonry MMC such as thin-joint technology has its place in the market and will not lead to deskilling of the masonry trades. Evidence from all avenues of the research confirms that if high skill levels are required, they are available.

What does raise concern is the fact that poor site conditions, training regimes and remuneration packages could, if left unchanged, markedly affect what the masonry sector has to offer in future years. A fully trained masonry tradesperson requires intelligence, athleticism and high levels of dexterous skill, but the wider construction industry continues to treat many of these professionals as untrained labourers who should be hired as casually and cheaply as possible.

Conditions on some sites have not improved since the nineteenth century and hardly serve to attract young people considering their career options. Some developers, while not evolving in terms of good site practice, have somehow managed to increase their profit margins. The time has surely come to grade developers in terms of levels of competence and their ability to provide product information to help building purchasers make decisions on how their money is being spent.

Furthermore, the fact that NVQs continue to effectively split trades or professions into a series of small competencies cannot be seen as a positive move, and the masonry sector would appear to be relying upon the individual drive and determination of some trainees to achieve the high levels of competence that are not necessarily required to become employed as a tradesperson. The time to re-examine the benefits of all-encompassing apprenticeship training schemes, which give college lecturers the time and resources required to produce first-class tradespeople, has surely arrived. Otherwise, the false claims of skills shortages now being made by advocates of MMC are likely to become a stark reality in the near future.

The construction industry now works at a pace primarily set by profit, not quality, with on-site mortar production being a victim of this approach. This is unfortunate in many respects. In this sense, it would seem that the way forward is the use of factory-produced mortar mixes of a known quality, as it is clear that site-based mortar production by hand is extremely difficult to quality assure. Materials such as paints, adhesives, grouts and sealants are not mixed on site, their quality is assured by the information available on their packaging. As a material essential to the success of a masonry building, it is very difficult to see why mortar should continue to be produced in an uncontrolled manner.

6.4.4 Enduring aesthetic appeal

However robust a building may be, it is less likely to survive the projected lifespan of its structure if its aesthetic appeal falls quickly out of fashion. Planning bodies, under increasing pressure to meet deadlines set by policy and large organisations, continue to in many cases approve bland 'one approach' designs, which in some instances are replicated on a nationwide basis. Such designs often give little or no thought to the use of local materials or the maintenance of a local vernacular. The loss of local identities is both sad from an aesthetics point of view and detrimental to the lifespan of some buildings, such as supermarkets.

Although it is extremely difficult to accurately measure the embodied energy levels involved in the transportation of mass-produced materials, it is safe to say that in comparison with building on the basis of the locality of materials and labour, the overall effect upon the environment will be a detrimental one.

6.4.5 Design for climate change

Taking all of the findings of this book into account, designing masonry for climate change is probably the biggest challenge that the sector faces, even though there is no doubt that plain masonry offers excellent thermal mass.

Again, all data suggest that planning bodies have much to answer for in terms of allowing (and indeed sometimes encouraging) developments that simply maximise density and give little consideration to a building's interaction with its environment. The Code for Sustainable Homes (which, as Dr Jacqui Glass commented, should have been the Code for Sustainable *Buildings*) will, hopefully, set the ball rolling towards the widespread consideration of the orientation of buildings, shading and passive solar design. Following what has recently been the warmest decade on record, it is hoped that the code will also be changed to recognise the need to design for climate change and not simply design to prevent it.

Placing insulation in a cavity jeopardises durability. Placing insulation on the outside of a building maximises thermal mass but limits the choice of external finishes to render or cladding. Such systems can also be susceptible to impact damage. Placing insulation on the internal face of external walls maximises future refurbishment potential and allows the use of an authentic faced masonry finish externally but results in a loss of thermal mass to the walls. However, the excellent load-bearing properties of masonry mean that, in this scenario, the utilisation of heavyweight intermediate flooring systems could serve to counter this.

In terms of wall construction, the answer to this conundrum may, in the end, be reliant upon a technological breakthrough rather than a composite masonry system such as that investigated by Hanson plc through Oxford Brookes University. Another option open to the masonry sector is to create masonry units with such high inherent thermal performance levels that no insulation would be required at all, such as the lime/hemp masonry system being developed by Lime Technology or the cellular clay units offered by German manufacturer Ziegel. Having already used the lime/hemp system on a number of developments, Ian Pritchett seems confident of its potential, but as a 'full wall cast on site' system and not in the more traditional mortar and masonry unit that has been previously used. Conversely, cellular clay units appear to be growing in popularity.

7 LOOKING FORWARD

Brick is not part of the fully automated, robotic world we all live in; it belongs to the more human side of construction – designed to fit in the hand, to be easily lifted and when in place to give a feel of permanence through the sheer mass of material. In contrast to the relatively flimsy walls of modernism, brick gives substance to a building.

Shuttleworth (2006, p. 5)

7.1 INTRODUCTION

From a position of little change over thousands of years, masonry has, for numerous reasons, had to evolve rapidly over the last 150 years or so. Since its widespread introduction in the 1940s, the cavity wall has proved to be problematic and has recently started to become a vessel for insulation rather than a simple masonry system in its own right. The time has come for the construction industry to pause and take stock; to continue along the road of cavities wider than 300 mm, filled with insulation with questionable durability, would seem to make very little sense. The construction client and industry in general have become increasingly interested in buildings for least cost, least time and least effort, rather than the quality and durability of the end product. To become more sustainable, attitudes must change, and quickly.

Overall, it would seem that the way forward is to take a step backwards. Large UK manufacturers are either advertising solid wall systems or have them in development, and manufacturers of solid wall systems from the continent have started to break into the UK market. This speaks volumes when considering the industry's confidence in the future of the cavity wall, as does the vast majority of the research data collected by the author over a four-year period. It seems a shame that companies manufacturing different types of masonry unit (eg clay bricks, calcium silicate bricks, dense concrete blocks, aircrete products, cellular clay blocks) continue to work in isolation from each other when each could so very easily complement the other. Solid wall construction has the ability to provide greater thermal mass than cavity wall construction, offers better levels of airtightness and allows far greater levels of buildability. Although more masonry units may be required in its construction, costs in comparison with cavity wall construction can be similar due to the myriad of accessories required by cavity walls. The extra materials required in solid wall construction are also likely to be offset by lower long-term maintenance needs and improved levels of durability and adaptability.

Manufacturers of masonry components are doing a great deal to reduce their impact on the environment, and in turn are reaping the rewards in terms of vastly reduced energy costs. On the grounds of an improved business profile and sound long-term financial planning, the vast amounts being spent on developments such as improved kiln technology are excellent investments, and although it is obviously in the interest of any manufacturer to reduce energy use, it appears that more vast amounts of money may need to be invested in R&D before further major advances are made.

The high levels of energy needed to manufacture the most common materials used to construct masonry buildings are often cited as the main reason that masonry is 'unsustainable'. However, the longer a masonry building lasts, the lower its effect will be upon the environment due to the spread of its embodied energy over time. If the building lasts for hundreds of years, its effect will be minimal in comparison with other methods of construction. This could be said to be the same for masonry buildings that fall out of fashion, but whose components can be readily taken apart and used again. Despite claims to the contrary, the masonry sector could do far more to promote new masonry that is subsequently built with future reclamation in mind. The government could do much to encourage future reclamation through incentives such as the introduction of an 'as-built' database to aid future generations of designers and demolition contractors, a far higher rate for landfill tax and a broad requirement for sustainability appraisals as part of the planning application process.

It is encouraging to see manufacturers utilising waste materials/recyclate in the makeup of their products, but disheartening to discover that planning bodies continue to stand in the way of attempts to dispose of waste by incineration through cement kilns, when this is surely a positive step in environmental terms.

Planning bodies must also take criticism for the fact that they do little to drive the need for buildings that interact with their local and natural environments and are equipped to cope with the onset of climate change. To date, not enough local authorities have had the foresight to introduce measures such as the requirement for sustainability statements. A closer working relationship with building control bodies would also serve to address such issues, which, if not carefully considered now, will create massive problems in the future.

In light of question marks against the future viability of cavity wall construction, a major problem presents itself: that of the positioning of insulation. Designers currently have a choice as to whether to render the outside of a building and therefore insulate externally, to provide a faced masonry finish to walls that would be insulated internally or to use materials with inherent thermal performance that require no additional insulation. Masonry needs to start being treated as an element in its own right and built with longevity and future adaptability in mind, thereby giving future generations the opportunity to upgrade buildings as necessary without the need for major structural works. Taking all of this into account, the following short statement by the author would seem to sum up the most sustainable model for low- to medium-rise sustainable masonry structures:

Sustainable masonry should ideally consist of a standardised system of solid wall construction built with locally sourced materials with the lowest possible levels of embodied energy. Masonry units should be manufactured to aid future reclamation and should be used in conjunction with a masonry mortar with a bond strength that will also do so. External walls should be built in a manner and with materials that will ensure long-term aesthetic appeal and durability. As a whole, a masonry structure should exhibit high levels of airtightness and be designed and built to optimise its orientation in order to maximise the natural conditioning of its internal environment and minimise the need for primary energy use. If used, insulants should be placed in an accessible location to facilitate future upgrade or adaptation of the masonry.

Such a system could consist of a clay or stone external face tied into a backing of more thermally acceptable masonry (aircrete, cellular clay or even lime/hemp), all bonded together with a low-strength masonry mortar, although the options seem endless. Dense solid concrete blocks, cellular clay blocks, cellular dense concrete blocks and jointless systems could all be mixed and matched through a number of combinations along with current and future insulation technology.

After many thousands of years as the primary and proved method used to construct buildings, the expanding array of manufacturing techniques and subsequent product ranges should ensure that masonry construction has a very bright future. However, a watershed has now been reached where masonry needs to evolve markedly to meet the needs of the twenty-first century and beyond.

7.2 FUTURE NEEDS
7.2.1 Energy consumption during manufacture

Manufacturers of masonry materials have done much in recent years, but can clearly go further in terms of lowering energy consumption and resultant CO_2 emissions. Further research should be carried out into possible kiln efficiency improvements and the use of alternative waste fuels. More research also needs to be carried out in order to provide accurate information on the levels of embodied energy in all construction products, which should then be displayed within trade literature and on packaging. The development of magnesium oxide cements could play a major part in the future of masonry construction.

7.2.2 Reclaimability

Manufacturers of masonry units should explore ways in which their products can be made to aid their future reclamation, and the masonry sector should promote the use of factory-mixed low-strength masonry mortars that will facilitate the future reclamation of masonry units. To encourage the use of reclaimed materials, the UK government should investigate the introduction of a national database of as-built information.

7.2.3 Regulation

Sustainability statements should widely become a requirement of the planning application process, and planning and building control bodies should investigate ways in which they can combine their expertise in order to ensure that proposed buildings will be energy efficient and able to cope with climate change.

7.2.4 Quality and craftsmanship

In terms of maintaining high levels of quality and workmanship, it is imperative that the training and treatment of masonry tradespeople improve markedly in the near future if masonry is to maintain its position as the foremost method of construction in the UK. Serious consideration should be given to the grading of contractors/developers on the grounds of site environment, quality of finished product and information provided on the ethos behind the construction of their buildings.

7.2.5 Evolution

The masonry sector should continue to explore modern solid wall forms of construction and preferably, manufacturers of different types of masonry unit should collaborate to ensure that optimum solutions are obtained.

REFERENCES

Abley I (2007). Forget the bogus idea of a skills shortage. Instead we should ask how much we will pay for quality. Construction Manager (March) p. 11.

Addis B (2006). Building with reclaimed components and materials: a design handbook for reuse and recycling. London, Earthscan.

Aircrete Bureau (2005). Factsheet 5 – the use of aircrete's solid wall construction. Sevenoaks, Aircrete Bureau.

Arup Research & Development (2005). UK housing and climate change: heavyweight vs. lightweight construction (C Twinn and J Hacker for Bill Dunster Architects). London, Ove Arup & Partners Ltd.

Ashurst J (1997). The technology and use of hydraulic lime. Accessed 23 November 2004. Available at: www.buildingconservation.com (Articles > Lime mortars and renders).

Ashurst J and Dimes F E (1990). Conservation of building and decorative stone, volume 1. London, Butterworth-Heinemann.

Baiche B, Kendrick C, Ogden R and Walliman N (2007). The development of a composite masonry unit for the UK construction industry. Masonry International (spring) 11–28.

BBC News (online) (a). 'Binding' carbon targets proposed. Accessed 14 March 2007. Available at: http://news.bbc.co.uk/1/hi/uk_politics/6444145.stm or log on to http://news.bbc.co.uk and search by article title.

BBC News (online) (b). 2007 to be 'warmest on record'. Accessed 18 March 2007. Available at: http://news.bbc.co.uk/1/hi/sci/tech/6228765.stm or log on to http://news.bbc.co.uk and search by article title.

BBC News (online) (c). Terrace houses lead property boom. Accessed 18 March 2007. Available at: http://news.bbc.co.uk/1/hi/business/6433835.stm or log on to http://news.bbc.co.uk and search by article title.

BBC Weather (online). 2006 – a year for the record books. Accessed 18 March 2007. www.bbc.co.uk/cgi-bin/education/betsie/parser.pl

Beare M (2004). Building commercially with lime mortar – justifying what we know will work – the engineering aspects. Journal of the Building Limes Forum, 11 15–22.

Beggs C (2002). Energy: management, supply and conservation. Oxford, Butterworth-Heinemann.

Bell J (2002). Doing your research project: a guide for first-time researchers in education and social science. Buckingham, Open University Press.

Bennet B (1999). Lime mortar aggregates and their effect on the performance of a lime mortar. Accessed 11 December 2004. Available at: www.thelimecentre.co.uk/aggregates.htm

Berge B (2000). The ecology of building materials. Oxford, Architectural Press.

Best R and de Valence G (2002). Design and construction: building in value. Oxford, Butterworth-Heinemann.

Blackman D (2007). Developers use appeals to blackmail the council. Building (27 July) pp. 40–41.

Blount B (1920). Cement. London, Longmans, Green & Co.

Bogue R H (1947). The chemistry of Portland cement. New York, Reinhold.

Bourke K, Ramdas V, Singh S, Green A, Crudgington A and Mootanah D (2005). Achieving whole life value in infrastructure and buildings. BRE Report BR 476. Bracknell, IHS BRE Press.

Bowyer B (1993). History of building. Builth Wells, Attic Books.

Brand S (1994). How buildings learn. New York, Penguin.

BRE (1991). Building mortar. Digest DG 362. Bracknell, IHS BRE Press.

BRE (2003). Construction and demolition waste. Good Building Guide GG 57, part 1. Bracknell, IHS BRE Press.

BRE (2005). Standard for innovative housing. Constructing the future (spring) p. 5.

Brick Development Association (1999). A sustainable strategy for the brick industry. Windsor, BDA.

Brick Development Association (2001). Use of traditional lime mortars in modern brickwork. Windsor, BDA.

Brick Development Association (2007). Bricklayer shortage. Brick bulletin (spring) p. 11.

Briggs M S (1925). A short history of the building crafts. Oxford, Clarendon Press.

British Cement Association (2009). Novel cements: low-energy, low-carbon cements. Camberley, BCA.

British Geological Survey (2005). Mineral profile: building and roofing stone. London, BGS.

British Standards Institution (1976). Specifications for building sands from natural sources. British Standard BS 1200. London, BSI.

British Standards Institution (1981). Specification for ready-mixed building mortars. British Standard BS 4721. London, BSI.

British Standards Institution (1985). Specification for clay and calcium silicate modular bricks. British Standard BS 6649. London, BSI.

British Standards Institution (1991). Code of practice for external renderings. British Standard BS 5262. London, BSI.

British Standards Institution (2000). Buildings and constructed assets – Service life planning – General principles. British Standard BS ISO 15686-1. London, BSI.

British Standards Institution (2001a). Code of practice for use of masonry – Materials and components, design and workmanship. British Standard BS 5628-3. London, BSI.

British Standards Institution (2001b). Building lime – Definitions, specifications and conformity criteria. British Standard BS EN 459-1. London, BSI.

British Standards Institution (2003a). Specification for mortar for masonry – Masonry mortar. British Standard BS EN 998-2. London, BSI.

British Standards Institution (2003b). Specification for masonry units – Calcium silicate masonry units. British Standard BS EN 771-2. London, BSI.

British Standards Institution (2003c). Specification for masonry units – Clay masonry units. British Standard BS EN 771-1. London, BSI.

British Standards Institution (2003d). Specification for masonry units – Manufactured stone masonry units. British Standard BS EN 771-5. London, BSI.

British Standards Institution (2003e). Specification for masonry units – Aggregate concrete masonry units. British Standard BS EN 771-3. London, BSI.

British Standards Institution (2003f). Specification for masonry units – Autoclaved aerated concrete masonry units. British Standard BS EN 771-4. London, BSI.

British Standards Institution (2005a). Code of practice for use of masonry – Structural use of unreinforced masonry. British Standard BS 5628-1. London, BSI.

British Standards Institution (2005b). Clay and calcium silicate bricks of special shapes and sizes – Recommendations. British Standard BS 4729. London, BSI.

British Standards Institution (2005c). Specification for masonry units – Natural stone masonry units. British Standard BS EN 771-6. London, BSI.

British Urban Regeneration Association (2005). Modern methods of construction: evolution or revolution? London, BURA.

Brundtland G (ed) (1987). Our common future: the world commission on environment and development (Brundtland Report). Oxford, Oxford University Press.

Campbell J W P and Pryce W (2003). Brick: a world history. London, Thames & Hudson Ltd.

Celcon (2005). Rå House (trade literature). Sevenoaks, Celcon.

CEMBUREAU (1998). Climate change, cement and the EU. Brussels, CEMBUREAU.

CEMBUREAU (1999). Best available techniques for the cement industry. Brussels, CEMBUREAU.

CERAM (2004). Traditional Plus – single and two storey design guide. Stoke-on-Trent, CERAM.

Chartered Institute of Building (2002). Sustainability and construction (CIOB information and guidance series). Ascot, CIOB.

Concrete Centre (2006). Thermal mass for housing: concrete solutions for the changing climate. Camberley, Concrete Centre.

Construction Industry Research and Information Association (1999a). The reclaimed and recycled construction materials handbook (C513). London, CIRIA.

Construction Industry Research and Information Association (1999b). Waste minimisation and recycling in construction – boardroom handbook. London, CIRIA.

Construction Industry Research and Information Association (1999c). Waste minimisation and recycling in construction (PR28). London, CIRIA.

Construction Industry Research and Information Association (2001). Demonstrating waste minimisation benefits in construction (C536). London, CIRIA.

Cuffe N (2003). Get it right: masonry walls. Building (31 October).

Davis Langdon (2007). Spon's architects' and builders' price book 2007. Abingdon, Taylor & Francis.

De Vekey R C (1993). Cavity walls – still a good solution. Conference proceedings No. 5, London, May 1993. London, British Masonry Society, pp. 35–38.

De Vekey R C (1999). Clay bricks and clay brick masonry. Digest DG 441, part 2. Bracknell, IHS BRE Press.

De Vekey R C, Tarr K and Worthy M (1989). Workmanship and the performance of wall ties: effect of depth of embedment. Conference proceedings No. 3, London, March 1989. London, British Masonry Society, pp. 74–77.

Department for Communities and Local Government (online) (2007). Extracts from a speech given by Yvette Cooper MP to The Home Builders' Federation Roundtable Summit on 9 January 2007. Accessed 13 February 2007. Available at: www.communities.gov.uk/speeches/corporate/zero-carbon

Department for Communities and Local Government (2007a). Building and roofing stone: mineral planning factsheet. London, DCLG.

Department for Communities and Local Government (2007b). Building a greener future: policy statement. London, DCLG.

Department of the Environment, Transport and the Regions (2000). Waste strategy 2000 for England and Wales, part 1. London, DETR.

Edwards B (1999). Sustainable architecture: European directives and building design. Oxford, Architectural Press.

Edwards B (2001). Rough guide to sustainability. London, RIBA Publications.

Energy Efficiency Best Practice Programme (2002). BedZED – Beddington Zero Energy Development, Sutton. London, Energy Efficiency Best Practice Programme.

English Heritage (1997). The English Heritage directory of building limes. Shaftesbury, Donhead.

EnviroCentre Ltd (2002). A report on the demolition protocol. London, EnviroCentre Ltd.

Environment Agency (2001). Integrated pollution prevention and control (IPPC) – guidance for the cement and lime sector. Bristol, Environment Agency.

Exall I (2007). A helping hand. Public sector and local government building (June) pp. 19–20.

Foresight Lime Research Team (2003). Hydraulic lime mortar for stone, brick and block masonry. Shaftesbury, Donhead.

Gervis J (2004). Sustainability – a way forward: marrying the old and new. Journal of the Building Limes Forum, 11 79–85.

Hammett M (1991). The role of brick in our environment. Windsor, BDA.

Hammond M (1990). Bricks and brick making. Oxford, Shire Publications Ltd.

Hammond G and Jones C (2006). Inventory of carbon and energy (ICE). Bath, University of Bath.

Harrison W H (1989). How traditional is brickwork? Conference proceedings No. 3, London, March 1989. London, British Masonry Society, pp. 66–68.

Harrison W H (1993). Avoiding latent mortar defects in masonry. Information Paper IP 10/93. Bracknell, IHS BRE Press.

Harrison J (2004). Carbonating and hydraulic mortars. Accessed 12 February 2005. Available at: www.tececo.com/files/conference_papers

Harrison H W and de Vekey R C (1998). BRE building elements: walls, windows and doors. BRE Report BR 352. Bracknell, IHS BRE Press.

Hazael P S (1993). Cavity and solid walls in aircrete masonry. Conference proceedings No. 5, London, May 1993. London, British Masonry Society, pp. 39–40.

Hendry A W and Khalaf F M (2001). Masonry wall construction. London, Spon Press.

Hewlett P C (ed) (2001). Lea's chemistry of cement and concrete. Oxford, Elsevier.

Highfield D (2000). Refurbishment and upgrading of buildings. London, E & FN Spon.

Historic Scotland (1998). Preparation and use of lime mortars. Edinburgh, Historic Scotland.

Hobbs G and Collins R (1997). Demonstration of reuse and recycling of materials: BRE energy-efficient office of the future. Information Paper IP 3/97. Bracknell, IHS BRE Press.

Hodgson G (2008). An introduction to PassivHaus: a guide for UK application. Information Paper IP 12/08. Bracknell, IHS BRE Press.

Hogg J, Roberts J and Fried A (2002). Prefabricated brickwork – what lessons can be learnt from other materials and industries. Conference proceedings No. 9, London, November 2002. London, British Masonry Society.

Holmes S and Wingate M (2002). Building with lime. London, ITDG Publishing.

House of Commons (online). Memorandum submitted by the British Cement Association. Accessed 12 February 2005. www.publications.parliament.uk/cgi-bin/ukparl

Housebuilder (online). Building tomorrow's world not today's quick fix (press release, 2 July 2004). Accessed 20 October 2006. Available at: www.housebuilder.org.uk/social (Press releases > 2004).

Howell J (1995). Technical and operational difficulties in cavity wall and solid wall construction. Conference proceedings No. 7, London, October 1995. London, British Masonry Society, pp. 172–174.

Hurley J and McGrath C (2001). Deconstruction and reuse of construction materials. BRE Report BR 418. Bracknell, IHS BRE Press.

IAM Motoring Trust (online). Static high petrol prices deprive high street shops of potential 120 million. Accessed 1 April 2007. Available at: www.iam.org.uk (Policy and research > News > News archive > 2005).

Ibstock Brick Ltd (2006a). 2005 environmental report. Ibstock, Ibstock Brick.

Ibstock Brick Ltd (2006b). Ecoterre earth bricks (trade literature). Ibstock, Ibstock Brick.

Institution of Structural Engineers (1999). Building for a sustainable future: construction without depletion. London, Institution of Structural Engineers.

Institution of Structural Engineers (2005). Manual for the design of plain masonry in building structures. London, Institution of Structural Engineers.

Irvine-Whitlock (2007). Bricklaying crisis? What crisis? Brick Works (p. 1). Windsor, Association of Brickwork Contractors.

Jones Sir D (2007). Out with the new. Building (23 February) pp. 34–35.

Kingspan (2000). Injected mineral fibre full fill cavity wall insulation, workmanship, voids and heat loss – a White Paper. Leominster, Kingspan.

Kingspan Insulated Panels UK (2006). Kingspan quick guide to the 2006 Building Regulations Approved Document Part L2 conservation of fuel and power (England & Wales). Holywell, Kingspan.

Kingspan Insulation Ltd (2007). Kooltherm K8 cavity board. Leominster, Kingspan.

Knauf Insulation (online). DriTherm cavity slabs. Accessed 20 August 2009. Available at: www.knaufinsulation.co.uk (Products > Glass mineral wool slabs > DriTherm cavity slabs).

Lane T (2006). A lot of the guys won't work on timber frame again. Building Magazine (1 December) pp. 22–25.

Lazarus N (2002). Construction materials report: toolkit for carbon neutral developments, part 1. Wallington, BioRegional Development Group.

Lazarus N (2005). Potential for reducing the environmental impact of construction materials. Wallington, BioRegional Development Group.

Livesey P (2002). Succeeding with hydraulic lime mortars. Journal of Architectural Conservation, 2 (July) 23–37.

Long J (1989). Solving mortar problems. National Builder (September).

McKay W B (1938, reprinted 1947). Building construction, volume 1. London, Longmans.

McKay W B (1944). Building construction, volume 2. London, Longmans.

McNeilly T (1993). Australian experience of nine-inch brickwork: watertight for 130 years. Conference proceedings No. 5, London, May 1993. London, British Masonry Society, pp. 41–44.

Mason J (1992). Solid masonry walls 1: future technology. Architects' Journal (3 June) pp. 47–55.

Modern Masonry Alliance (2006). MMA response to the ODPM Code for Sustainable Homes consultation. Leicester, MMA.

MORI (2001). Attitudes towards house construction. London, MORI.

Mortar Industry Association (1988). A basic guide to brickwork mortars, part 1: materials, mixes and selection. London, Mortar Industry Association.

Mortar Industry Association (2004a). Learning text – introduction to mortar. London, Mortar Industry Association.

Mortar Industry Association (2004b). CPD course notes – design and technology of masonry mortar. London, Mortar Industry Association.

Mortar Industry Association (2004c). Learning text – properties of masonry mortar. London, Mortar Industry Association.

Morton T (1986). Designing for movement in brickwork (BDA Design Note 10). Windsor, BDA.

National Audit Office (2005). Using modern methods of construction to build homes more quickly and efficiently. London, National Audit Office.

NHBC Foundation (2008). The use of lime-based mortars in new build. Bracknell, IHS BRE Press.

Oates J A M (1998). Lime and limestone: chemistry and technology, production and uses. Weinheim, Wiley-VCH.

Office of the Deputy Prime Minister (2004a). Planning for the supply of natural building and roofing stone in England and Wales: a summary. London, ODPM.

Office of the Deputy Prime Minister (2004b). Planning for the supply of natural building and roofing stone in England and Wales. London, ODPM.

Office of the Deputy Prime Minister (2004c). Approved document C: site preparation and resistance to contaminants and moisture. London, TSO.

Office of the Deputy Prime Minister (2006). Approved document L1B: conservation of fuel and power in existing dwellings. London, TSO.

Oxley R (2003). Survey and repair of traditional buildings – a sustainable approach. Shaftesbury, Donhead.

Parkinson G, Shaw G and Beck J (1996). Appraisal and repair of masonry. London, Thomas Telford Ltd.

Permarock (2006). Phenolic (trade literature). Loughborough, Permarock.

Pritchett I and Dukes K (2003). The sustainability of lime in new buildings. Accessed 12 February 2005. www.limetechnology.co.uk/press02.html

Regenerate (2006). What makes a building sustainable? Building (21 July).

Ribar J W and Dubovoy V S (1988). Investigation of masonry bond and surface profile of brick. Masonry: Materials, Design, Construction and Maintenance. (Ed H A Harris.) Philadelphia, ASTM.

Roaf S (2004). Closing the loop: benchmarks for sustainable buildings. London, RIBA Enterprises Ltd.

Roaf S, Fuentes M and Thomas S (2004). Ecohouse 2: a design guide. Oxford, Architectural Press.

Roberts J J (1993). Solid walls v cavity walls – some observations. Conference proceedings No. 5, London, May 1993. London, British Masonry Society, pp. 31–34.

Roberts J J (1998). Sustainable masonry construction. Conference proceedings No. 8, London, November 1998. London, British Masonry Society, pp. 1–5.

Royal Institute of British Architects (1934). Modern practice in brickwork. Journal of the Royal Institute of British Architects, 61 (July) 858–859.

Sear L (2007). PFA – the environment and sustainability. Building Engineer (September) pp. 12–13.

Setra Marketing Ltd (2001). Information Sheets. Accessed 11 March 2005. Available at: www.stastier.co.uk

Shuttleworth K (2006). It all stacks up. Brick bulletin (spring) p. 5.

SIG Insulations (2009). A guide to sustainable insulation materials. Sheffield, SIG Insulations.

Smith A S (1999). Look before you leap. Brick bulletin (summer) pp. 29-31.

Smith P F (2006). Architecture in a climate of change – a guide to sustainable design. Oxford, Architectural Press.

Smith M, Whitelegg J and Williams N (1998). Greening the built environment. London, Earthscan Publications Ltd.

Spence R, Macmillan S and Kirby P (eds) (2001). Interdisciplinary design in practice. London, Thomas Telford.

Stern Sir N (2007). The economics of climate change: the Stern Review. Cambridge, Cambridge University Press.

Sutherland R J M (1993). Have we got it all wrong? Conference proceedings No. 5, London, May 1993. London, British Masonry Society, pp. 1–8.

Thermalite (2005). Thin joint masonry. Birmingham, Thermalite.

Thomas K (1989). Workmanship defects in the installation of wall ties: effect of depth of embedment. Conference proceedings No. 3, London, March 1989. London, British Masonry Society, pp. 78–80.

Town and Country Planning Association (2007). Climate change: adaptation by design. London, Town and Country Planning Association.

Tutt J N (1990). Site control of mortar manufacture. London, British Masonry Society.

Vicat L J (1837, reprinted 2003). Mortars and cements. Shaftesbury, Donhead.

Viridis (2002). The construction industry mass balance: resource use, wastes and emissions. London, Viridis.

Wingate M (1987). An introduction to building limes. London, SPAB.

Woolley T, Kimmins S, Harrison P and Harrison R (1997). Green building handbook, volume 1. London, Spon Press.

XCO2 (2002). Insulation for sustainability – a guide. London, XCO2.

Yool A I G and Lees T P (1995). A survey of UK building sands. Conference proceedings No. 7, London, October 1995. London, British Masonry Society, pp. 59–62.

USEFUL CONTACTS

Association of Brickwork
Contractors
26 Store Street
London WC1E 7BT
www.brickworkcontractors.info

BRE
Bucknalls Lane
Watford
Hertfordshire WD25 9XX
www.bre.co.uk

Brick Development Association
Woodside House
Winkfield
Windsor
Berkshire SL4 2DX
www.brick.org.uk

British Cement Association
Riverside House
4 Meadows Business Park
Station Approach
Blackwater
Camberley
Surrey GU17 9AB
www.cementindustry.co.uk

British Geological Survey
Natural History Museum
Cromwell Road
London SW7 5BD
www.bgs.ac.uk

British Precast
60 Charles Street
Leicester LE1 1FB
www.britishprecast.org

Building Limes Forum
Glasite Meeting House
33 Barony Street
Edinburgh EH3 6NX
www.buildinglimesforum.org.uk

Celotex Ltd
Lady Lane Industrial Estate
Hadleigh
Ipswich
Suffolk IP7 6BA
www.celotex.co.uk

CERAM
Queens Road
Penkhull
Stoke-on-Trent
Staffordshire ST4 7LQ
www.ceram.co.uk

The Concrete Society
Riverside House
4 Meadows Business Park
Station Approach
Blackwater
Camberley
Surrey GU17 9AB
www.concrete.org.uk

Donhead Publishing Ltd
Lower Combe
Donhead St Mary
Shaftesbury
Dorset SP7 9LY
www.donhead.com

Esk Building Products
Brisco
Carlisle
Cumbria CA4 0QY
www.eskbuildingproducts.com

Furness Brick & Tile Co Ltd
Dalton Road
Askam-in-Furness
Cumbria LA16 7HF
www.furnessbrick.co.uk

H+H Celcon UK Ltd
Head Office
Celcon House
Ightham
Sevenoaks
Kent TN15 9HZ
www.hhcelcon.co.uk

Ibstock Brick Ltd
Head Office
Leicester Road
Ibstock
Leicestershire LE67 6HS
www.ibstock.com

International Masonry Society
Shermanbury
6 Church Road
Whyteleafe
Surrey CR3 0AR
www.masonry.org.uk

Irvine-Whitlock Ltd
Brickstone House
Priory Business Park
Bedford MK44 3JW
www.irvinewhitlock.co.uk

Kingspan Insulation Ltd
Pembridge
Leominster
Herefordshire HR6 9LA
www.insulation.kingspan.com

Knauf Insulation Ltd
PO Box 10
Stafford Road
St Helens
Merseyside WA10 3NS
www.knaufinsulation.co.uk

Lime Technology Ltd
Unit 126
Milton Park
Abingdon OX14 4SA
www.limetechnology.co.uk

Loughborough University
Department of Civil and Building Engineering
Loughborough
Leicestershire LE11 3TU
www.lboro.ac.uk/departments/cv

Mineral Products Association
Gillingham House
38–44 Gillingham Street
London SW1V 1HU
www.mineralproducts.org

Modern Masonry Alliance
4th Floor
60 Charles Street
Leicester LE1 1FB
www.modernmasonry.co.uk

Mortar Industry Association
Gillingham House
38–44 Gillingham Street
London SW1V 1HU
www.mortar.org.uk

Oxford Brookes University
School of the Built Environment
Headington Campus
Gipsy Lane
Oxford OX3 0BP
www.brookes.ac.uk/schools/be

Rockwool Ltd
Pencoed
Bridgend
South Wales CF35 6NY
www.rockwool.co.uk

Setra Marketing Ltd (importers of St Astier natural hydraulic limes)
Flat 1
Admirals Point
East Cowes
Isle of Wight PO32 6AH
www.stastier.co.uk

Sheffield Insulations
Hillsborough Works
Langsett Road
Sheffield S6 2LW
www.sheffins.co.uk

Stone Federation Great Britain
Channel Business Centre
Ingles Manor
Castle Hill Avenue
Folkestone
Kent CT20 2RD
www.stone-federationgb.org.uk

Thomas Armstrong (Concrete Blocks) Ltd
Bridge Road
Brompton-on-Swale
Richmond DL10 7HW
www.thomasarmstrong.co.uk

University of Bath
Department of Architecture and Civil Engineering
Bath BA2 7AY
www.bath.ac.uk/ace

University of Cambridge
Department of Architecture
1–5 Scroope Terrace
Cambridge CB2 1PX
www.arct.cam.ac.uk

Ziegel UK
Bennacott Barn
Boyton
Launceston
Cornwall PL15 8NR
www.ziegeluk.co.uk

INDEX

Page numbers in *italics* indicate illustrations

A
Abley, Ian 89, 90, 92, 96, 100, 102, 104, 106, 108, 112, 114
aesthetics, vernacular buildings 79–81, 121–2
aggregates 31–2, 77
aircrete blocks
 construction with 55
 durability 95
 production 42–5, *42*, *43*, *44*, 98
Aqua Claudia, Rome 36
ashlar 49, 80
Aspdin, Joseph 6

B
Bath, The Circus 65
Beaufort Park, London, fire 17, *17*
BedZED development, Sutton 83, *83*
blockwork, modern methods 52, 80, *80*
Brand, S 71
Brick Awards 74, *74*, *116*
bricks
 calcium silicate bricks 39–41, *40*, *41*, 55, 118
 clay bricks 34–9, *35*, *36*, *37*, *39*, 99
 concrete bricks/blocks 41–5, *42*, *43*, *44*, 55
 earth bricks 38, 39, *39*
 environmentally friendly 38, 118–19
 history 6–7
 production 9–10, *9*, *10*, 35–8, *37*, 99
 reclaimed 39, 58, 61
brickwork
 aesthetics 79, *79*
 craftsmanship 73–4, *74*
British Standards
 BS 1200 32
 BS 3921 34
 BS 4721 22
 BS 4729 34
 BS 5628 9, 60
 BS 5628-1 9, 22
 BS 5628-3 23, 34, 70, 72, 78
 BS 6073-1 34
 BS 6457 34
 BS 6649 34
 BS 18799 34
 BS EN 459-1 24
 BS EN 998-2 33
 BS ISO 15686-1 73
Brundtland Report 13
Budapest, Hungary, passive solar design 83
building layers, six S's 71, *71*
Building Regulations, Part L 12, 52, 68, 81, 104
Burdett, Michael 89, 108, 110

C
calcium silicate bricks 39–41, *40*, *41*, 55, 118
Calpin, Kevin 89, 108, 110
carbon dioxide
 embodied 20–1
 emissions 2–3, 26, 30–1, 54, 119
Castle Cement 30
cavity walls
 aesthetics 80, *80*
 construction 10–12, *11*
 durability problems 64–6, 84, 120, 123
 insulation 52–4, 119, 120
cellular clay units *12*, 52, 70, 122
cellular plastic, insulation 52, 53, *53*
cement
 carbon emissions 2, 54
 energy consumption 28–9, 54–5, 98
 history 8–9
 production 27–31, *29*, *30*, 99, 118
 strength of 59–61
Cemex 29
CERAM, walling system 66, *67*, 70, 84
Clarke, Martin 89, 90, 92, 94, 96, 100, 102, 104, 106, 108, 112
clay bricks 34–9, *35*, *36*, *37*, *39*, 99
clay grinding mill 9, *9*
climate change
 construction impacts 2–3
 construction requirements 14–16
 designing for 81–3, *82*, *83*, *84*, *85*, 122
Cockermouth, reclaimed building materials 62
Code for Sustainable Homes 122
composite masonry units 68, *68*, 122

concrete bricks/blocks 41–5, *42, 43, 44*, 98
construction industry
　environmental impacts 1–3, 14–16, 19–21, 124
　reclaimed materials 19–20, 55, 57–64, 84, 100–1, 119–20
　skills needed 73–5, 85, 108–9, 120–1
　views on sustainable masonry 87–115
Cotswold stone 86
craftsmanship
　traditional building methods 73–5, *74, 75*, 108–9, 120–1
　training 110–11, 126

D

demolition, reuse of materials 57–64, *62*
design
　for climate change 81–3, *82, 83, 84*, 85, 122
　for future scenarios 71–2, 119–20
Driver, Michael 89, 90, 92, 94, 96, 100, 102, 104, 106, 112, 114

E

earth bricks/blocks 38, 39, *39*
Ecoterre earth bricks/blocks 38, 39, *39*
Eddystone Lighthouse 6
Edgell, Geoff 89, 90, 92, 94, 96, 100, 102, 104, 106, 112
embodied energy, masonry materials 20–1, 54, 98–9, 119, 125–6
energy, use in product manufacturing 98–9, 118–19, 124, 125–6
English bond, brickwork *11*, 79, *79*
environmental impacts
　brick production 38, 99, 118–19

construction industry 1–3, 14–16, 19–21
expansion joints 71–2

F

fire risks 17, *17*
Flemish bond, brickwork 79, *79*
flintlime bricks *see* calcium silicate bricks
Foresight Lime Research Team 78
Fudge, Cliff 89, 90, 92, 94, 98, 100, 102, 104, 106, 109, 112, 118
Furness Brick 37

G

Glass, Jacqui 89, 90, 92, 94, 96, 98, 100, 102, 112, 114, 118
government support
　modern methods of construction (MMC) 15, 72–3, 92–3, 117
　sustainable construction 112–13, 124
Grange Quarry, Cumbria *47, 48*

H

Haberdashers' Hall, London 23, 72
Hanson plc, composite masonry units 68, *68*, 122
High Grange Developments 75, *75*
Hills, David 89, 90, 92, 94, 99, 101, 102, 105, 107, 113
hydraulic lime mortar
　aggregates 31–2
　classification 25
　history 6, 9
　production 24–7
　quality of 78
　use of 22–3, 72

I

Ibstock Brick Ltd, Ecoterre earth bricks/blocks 38, 39, *39*
Industrial Revolution 6

insulation
　cavity walls 52–4, 119, 120
　cellular plastic 52, 53, *53*
　internal or external 122, 124–5
　long-term performance 119
　mineral fibre 52, 53, *53*, 119
　solid walls 81, *82*, 122
　thermal efficiency 51–2

J

Jerwood Centre, Grasmere *74*

K

Kelly, Ruth 14
kilns, cement production 8, 28–9, *29*, 118

L

landfill tax 63
Lanercost Priory, Cumbria *51*
legislation, planning 114–15, 126
Lepenski Vir, Yugoslavia 5
lifespan, masonry buildings 102–3, 120
lime cycle 24
lime mortar
　history 5–6, 8–9
　see also hydraulic lime mortar
lime/hemp masonry sytem 122
limestone, building material 46
local materials
　bricks 61, 79
　stone 50, 55, 80
　use of 54, 84, 121–2

M

magnesium oxide cements 31, 126
masonry construction
　advantages 15–17
　environmental impacts 19–21, 124
　lifespan 102–3, 120
masonry materials, embodied energy 20–1, 54, 98–9, 119, 125–6

masonry units
  British Standards 34
  calcium silicate bricks 39–41, *40*, *41*, 55, 118
  clay bricks 34–9, *35*, *36*, *37*, 39
  concrete bricks/blocks 41–5, *42*, *43*, *44*, 98
  functionality 34
  insulation 51–4, *53*
  new technology 118–19
  stone 45–50, *47*, *48*, *49*, *51*
medieval brickwork 6–7, *7*
Michaëlis, Wilhelm 39
mineral fibre, insulation 52, 53, *53*
MMC *see* modern methods of construction
modern methods of construction
  drawbacks 15–17, *16*, *17*, 85, 117–18
  government promotion 15, 72–3, 92–3, 117
  masonry industry 94–5
  skills needed 73–5, 120–1
  whole life costs (WLC) 72–3, 96–7, 120
Monroe, Paul 89, 107, 109, 111
mortar
  aggregates 31–2
  cement production 27–31, *29*, *30*, 98
  factory-produced mixes 32–3, *33*, 78, 121, 126
  history 5–6
  hydraulic lime 6, 9, 22–7, 72, 78
  movement of 72
  quality of 76–8
  reclaimability 59–61, 84, 119–20, 126
  requirements 22–3
  sustainability 21–2

N

Natural Stone Awards 74, *74*, *86*
Netherlands, recycling 63
new housing, cavity walls 52–4
Novacem 31

P

passive solar design 83, *84*, 85
PassivHaus 52–3
PFA (pulverised fuel ash)
  aircrete block production 43, 45, 55
  hydraulic lime production 25
planning legislation 114–15, 126
Portland cement
  development of 6, 8–9
  production 27–31
  strength of 59–61
pozzolanic cement 6
prefabricated building systems, drawbacks 15–17, *16*, *17*
prefabrication, masonry units 74, 106–7
Pritchett, Ian 23, 89, 90, 93, 94, 97, 101, 103, 105, 107, 109, 113, 122
Profuel 30–1

Q

quarrying 46, *47*, *48*, 49

R

Rå House 42
Ransome, Frederick 8, 28
reclaimed materials 19–20, 55, 84, 100–1
  bricks 39, 58, 61
  certification 63–4
  mortar 59–61, 84, 119–20
  planning for 63–4, 119–20, 124, 126
  use in new buildings 57–64, *62*
recycling *see* reclaimed materials
renovation, solid wall buildings 68–9, *69*, 71
Roaf, Sue 89, 91, 93, 94, 97, 101, 103, 113, 115
Romans
  bricks 36, *36*
  construction methods 5–6
RSPB headquarters, Bedfordshire 23, 72

S

Sainsbury's supermarket, Cockermouth 62
St Astier Limes 29
St Pancras Station, London 74
sandlime bricks *see* calcium silicate bricks
sandstone 46, *47*, *48*, 49
'scenario buffered buildings' 71, 120
site conditions 75, 106–7, 121
six S's, building layers 71, *71*
skills needed, construction industry 73–5, 85, 108–9, 120–1
slate, as building material 74
Smeaton, Joseph 6
solar design, passive 83, *84*
solid walls
  adaptation 68–9, *69*, 125
  durability 64, *65*, 68–72
  insulation 81, *82*, 122
  'layers of change' 71
  modern methods 12, *12*, 70–1, 104–5, 123–4
  movement of 71–2
  recommended thicknesses 70
  traditional 10, *11*
  water penetration 70
Stern Review 14
stone
  aesthetics 80, *80*
  as building material 45–6, 49–50, *51*, 55, 74
  decay of 76
  imported 50
  preparation 46, *48*, 49, *49*
  quarrying 46, *47*, *48*, 49
stretcher bond 85
  brickwork 79, *79*
supermarkets, lifespan 1, 61, 79
sustainable construction, government support 112–13, 124
sustainable development 13
sustainable masonry
  definition 13–14, 90–1
  future of 125–6

sustainable masonry *(cont'd)*
  views of construction industry 87–115

T

terraced housing, energy efficiency 82–3, *83*
thermal efficiency, insulation 51–2
thermal mass, masonry buildings 81
traditional building methods
  craftsmanship 73–5, *74*, *75*, 108–9, 120–1, 126
  solid walls 10, *11*
  training 110–11, 126
Traditional Plus, single-skin masonry 66, *67*, 84
training, traditional building methods 110–11, 126

U

United States, cement mortars 59–60
University of Cambridge
  Clare College, Gillespie Centre *18*
  Queens' College, The Court *7*

V

Van Derburg 39
vernacular buildings, aesthetics 79–81, 121–2
Vicat, Louis 24
Vitruvius 5–6

W

wall ties, corrosion 65, 66
waste
  construction industry 1–2
  use in aircrete block production 43, 45, 55, 118
  use in clay brick production 38
  use in hydraulic lime production 25
  use in masonry products 118
whole life costs (WLC) 72–3, 96–7, 120
Winscales Moor, Cumbria, pit building conversion 68–9, *69*

Y

York Minster, stone working *48*, *49*
York St John University, De Grey Court *116*

Z

zero-carbon buildings, cavity walls 52, 56
Ziegel, cellular clay units *12*, 52, 70, 122

OTHER TITLES FROM IHS BRE PRESS

HEMP LIME CONSTRUCTION
A guide to building with hemp lime composites
The first complete guide to building with hemp lime

- Fully illustrated with numerous examples and design details
- Enables practical construction with this sustainable material
- Written with input from materials, design and technical specialists

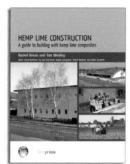

Hemp lime is a composite construction material that can be used for walls, insulation of roofs and floors and as part of timber-framed buildings. It provides very good thermal and acoustic performance, and offers a genuinely zero-carbon contribution to sustainable construction. Comprehensive guidance on using this novel material for housing and low-rise buildings is given for the first time in this illustrated book, which is full of practical information on materials, design and construction.

Ref: EP 85, ISBN: 978-1-84806-033-3, 2008

EARTH MASONRY
Design and construction guidelines
An inspirational introduction and guide to sustainable earth building

- Detailed guidance on contemporary construction using earth masonry
- Provides clients, designers and builders with clear, up-to-date information and advice
- More than 210 colour photos and diagrams

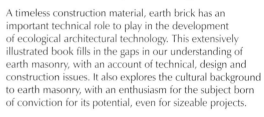

A timeless construction material, earth brick has an important technical role to play in the development of ecological architectural technology. This extensively illustrated book fills in the gaps in our understanding of earth masonry, with an account of technical, design and construction issues. It also explores the cultural background to earth masonry, with an enthusiasm for the subject born of conviction for its potential, even for sizeable projects.

Ref: EP 80, ISBN: 978-1-86081-978-0, 2008

EARTHSHIPS
Building a zero-carbon future for homes
What earthships are, why they have a low environmental impact and how to build them

- The most wide-ranging and up-to-date book on earthships
- Detailed guidance on designing and building earthships, based on first-hand experience
- Foreword by Kevin McCloud

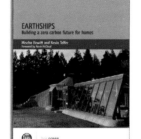

What exactly is an earthship? This book tells you everything you need to know to answer that question, and more. The earthship is a building concept that has evolved over the last 30 years, and represents a pioneering form of zero-carbon residential building that tackles a variety of sustainability issues. This book charts the building of the first earthships in the UK and their relevance to home building and architecture generally, as well as offering lessons about sustainable architecture and the legislative and regulatory culture that affects their construction.

Ref: EP 78, ISBN: 978-1-86081-972-8, 2007

RAMMED EARTH
Design and construction guidelines
Comprehensive guidance on rammed earth construction

- The most authoritative guide on rammed earth
- Fully illustrated with photos, diagrams and design details
- Brings together technical guidance with design and construction experience

Until now there has been no authoritative guidance on the use of rammed earth in the UK. This book presents state-of-the-art practical guidance on material selection, construction, structural design, architectural detailing, maintenance and repair of rammed earth. The aim of the book is to inform, develop and encourage the use of rammed earth wall construction for housing and other low- and medium-rise buildings. The guidance has been derived from extensive government-funded testing and research at the University of Bath.

Ref: EP 62, ISBN: 978-1-86081-734-2, 2005

THE GREEN GUIDE TO SPECIFICATION
Fourth edition
Guidance on making the best environmental choices when selecting construction materials and components

- Easy-to-use A+ to E graded system for each specification to enable comparison between similar materials and components
- Robust information based on a rigorous process of research, data collection and analysis
- An essential tool for all those seeking to reduce the environmental impact of their buildings
- Contains more than 1200 specifications

The fourth edition of *The green guide to specification* has been revised and updated by BRE to provide designers and specifiers with easy-to-use guidance on making the best environmental choices when selecting construction materials and components. It is more comprehensive than its predecessors and contains more than 1200 specifications used in six generic building types: commercial, educational, healthcare, retail, residential and industrial.

Ref: BR 501, ISBN: 978-1-84806-071-5, 2009

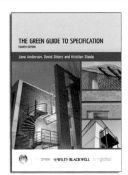

GREEN ROOFS AND FAÇADES
The case for making buildings greener

- Green roofs are part of the wider green and sustainable development agenda
- Provides accessible guide to green roofs
- Meets the rapidly growing interest in green roofs
- Numerous illustrations of green roofs and façades from around the world

Green roofs and façades on buildings offer a wide range of benefits, including attenuation of rainwater run-off, improved thermal stability and energy conservation, enhanced air quality, wildlife habitat and open space. This book provides an accessible overview of the development of green roofs and the contribution they can make to sustainable development. It explains the benefits of their use, and identifies the key aspects that must be considered in designing, building and maintaining them. It is fully illustrated with numerous examples of successful applications from around the world.

Ref: EP 74, ISBN: 978-1-86081-940-7, 2006

CONSERVATION AND CLEANING OF MASONRY
Part 1: Stonework
Part 2: Brickwork, blockwork and terracotta
Part 3: Renders, plasters and stucco
Companion Digest to DG 502

The use of appropriate conservation techniques is critical to the preservation of buildings. This three-part Digest gives advice on how to apply safe, effective, appropriate and enduring conservation techniques to a range of masonry materials. It outlines how to identify the material and associated mortars, and how to diagnose the causes of soiling and deterioration. It advocates the avoidance of cleaning processes in order to optimise the life of masonry while giving an outline guide to their use where acceptable.

Ref: DG 508, part 1, ISBN: 978-1-84806-061-6, 2008
Ref: DG 508, part 2, ISBN: 978-1-84806-113-2, 2009
Ref: DG 508, part 3, forthcoming publication

PRINCIPLES OF MASONRY CONSERVATION MANAGEMENT
Companion Digest to DG 508

This Digest provides an overview of the principles of the conservation management of masonry in historic buildings. It discusses the legislative framework in the UK (including interactions with planning authorities and conservation bodies, and regional variations in conservation policy) and the general principles of conservation practice, including minimal intervention, protection against disaster and neglect, use of identical replacement materials or the nearest analogues, respecting the spirit of the core of the building, maintaining a log of works, and agreeing a protocol for project management before a project begins. An Annex lists key events in the development of building conservation bodies, practice and legislation in the UK.

Ref: DG 502, ISBN: 978-1-86081-966-7, 2007

To place an order: tel: +44 (0)1344 328038 email: brepress@ihs.com web: www.brebookshop.com